I0064141

# Explaining Physics
## With the
## Help of Toys

# Explaining
# Physics
## With the
# Help of Toys

## Igor Kulaga
Kurchatov Institute, Russia

**World Scientific**

NEW JERSEY · LONDON · SINGAPORE · BEIJING · SHANGHAI · HONG KONG · TAIPEI · CHENNAI · TOKYO

*Published by*

World Scientific Publishing Co. Pte. Ltd.

5 Toh Tuck Link, Singapore 596224

*USA office:* 27 Warren Street, Suite 401-402, Hackensack, NJ 07601

*UK office:* 57 Shelton Street, Covent Garden, London WC2H 9HE

Library of Congress Control Number: 2022939240

**British Library Cataloguing-in-Publication Data**
A catalogue record for this book is available from the British Library.

EXPLAINING PHYSICS WITH THE HELP OF TOYS

Copyright © 2022 by World Scientific Publishing Co. Pte. Ltd.

*All rights reserved. This book, or parts thereof, may not be reproduced in any form or by any means, electronic or mechanical, including photocopying, recording or any information storage and retrieval system now known or to be invented, without written permission from the publisher.*

For photocopying of material in this volume, please pay a copying fee through the Copyright Clearance Center, Inc., 222 Rosewood Drive, Danvers, MA 01923, USA. In this case permission to photocopy is not required from the publisher.

ISBN 978-981-125-600-4 (hardcover)
ISBN 978-981-125-601-1 (ebook for institutions)
ISBN 978-981-125-602-8 (ebook for individuals)

For any available supplementary material, please visit
https://www.worldscientific.com/worldscibooks/10.1142/12827#t=suppl

Desk Editor: Joseph Ang

Typeset by Stallion Press
Email: enquiries@stallionpress.com

# To the Reader

Few would challenge Aristotle's idea that "knowledge begins with surprise". With little effort one can turn their memories back to their childhood, feeling the wonder of first encountering a new and unfamiliar toy. These feelings would arise from the oddity of this toy, which was confronting us with the laws of nature in disguise. Today, toys such as these are often classified as "scientific toys". They are interesting because of how they disguise the laws from which they were designed, the ways which allow them to produce their effects, and the purposes for which they were created. This purpose often lies in the mystery of their mechanism; just by using them, our research into this mystery has already begun.

As such toys are designed with the learning process in mind, it is a shame to squander their potential by leaving them to dissipate amongst our other childhood memories. Scientific toys are particularly suited to conveying ideas of a variety of physical laws, whether they are simple, novel, obscure, or complicated.

Sokolov Nikolaj Nikolajevich, a professor at the Russian State Geological Exploration University, was a pioneer, possibly the pioneer, of a method of scholastic education which incorporated such toys as demonstrations of their principles of design, and used toys in their classes in Russia during the 1970s. Their interactive lectures and demonstrations attracted large audiences to the Moscow Polytechnic Museum and afforded them a great reputation. To this day, Sokolov's methods influence the pedagogical procedures of progressive teachers of physics all over the world.

Scientific toys demonstrate ideas of the laws of physics from a multitude of topics. In any good catalogue of toys, one can find scientific toys; the range of such toys is ever increasing. Nonetheless, advertising materials often obscure the educational applications of particular toys. This book aspires to work towards correcting that situation. Within this book are four dozen different descriptions of toys relating to their domain in physics, in addition to materials on the technical and scientific properties of these toys and explorations of the history of discoveries and inventions behind that toy. All toys described within are manufactured industrially, thus one should not have great difficulty to find examples. In addition to these toys, the same stores and catalogues may have other scientific demonstration kits of varying complexities relating to chemistry, biology, and other natural sciences.

The virtues of scientific toys are numerous, however they are not the exclusive source of accessible demonstrations of particular principles from physics. There is no "official" character to the toys recommended which make them superior to another toy which demonstrates the same principles. Even a Barbie Doll, which at first glance might be thought to have nothing to do with physics, can help draw attention to physics. One might start a story from the topic of the simplest mechanisms in mechanics with the example of the hinges in the joints of the Barbie Doll.

Conversely, some toys will better demonstrate a difficult physical law or phenomenon than others and will serve better to arouse interest or be an effective visual aid. For example, quantum mechanics is quite an abstract science which resists obvious visual aids. However, by throwing two neodymium magnetic balls into the air, the way that children often use them, one could describe how the force of their attraction is based on the movement of electrons, already helping demonstrate electromagnetic phenomena in a tangible way which can be related back to the standpoint of quantum mechanics, while also being accessible and entertaining. Nonetheless, one must not get carried away with intricacies and turn these tools into distraction from assistance.

It is strange that purchased toys often tend to break quickly; in childhood many of us make toys from what we can access. This fact can be used with considerable benefit in the educational process. Nowadays on the internet one can find a whole host of recordings, which show the process of making a wide variety of toys for oneself, often from the most seemingly disposal of resources. This book provides a step-by-step description of the creation of a dozen homemade scientific toys.

It is crucial that in the process of working together on these toys and their subsequent tests, teachers will be able to tell children a lot about the physical principles which the toy demonstrates and the history of the invention and its conceptualization, amongst other things. Discovering together in a creative environment will leave a lasting impact on one's students.

Igor Kulaga
(Translated by F.P. Lewis)

# Contents

# Chapter 1

# Mechanics

## 1.1 "Stubborn Giraffe" (Pertaining to *Statics and Construction Mechanics*)

The stylised giraffe figure shown in Figure 1.1 can be attributed to a class of toys that are known as puppets. The history of the appearance of marionettes dates to the times of Ancient Egypt and Ancient Greece, where puppets were known to be used within theatre, driven by a puppeteer with the help of ropes, wooden sticks, or metal wires. The name for a manipulable puppet, a marionette, appeared much later. In the Middle Ages, people were very fond of dolls depicting the Virgin Mary, which were often given diminutive titles. This is one possible etymological root, though another proposed root was that this name, sounding like a diminutive of Mary, was given to dolls in Italy for the name of a puppet master from the sixteenth century who was named Marioni.

"Stubborn giraffe" has a very significant difference from other traditional puppets. With theatrical puppets, most of their threads hang loose when the limbs of the puppet are loose, and the puppeteer brings their puppets to life by playing with the strings. The toy giraffe's legs and neck are beads which have been strung together, the springs are made taut by a compressible spring hidden in the hollow base of the giraffe's platform. Unless the spring is compressed by the floor of the platform being pushed in on itself, the giraffe will remain upright. This unusual puppet can be changed from upright, when the spring is left uncompressed, to a state of

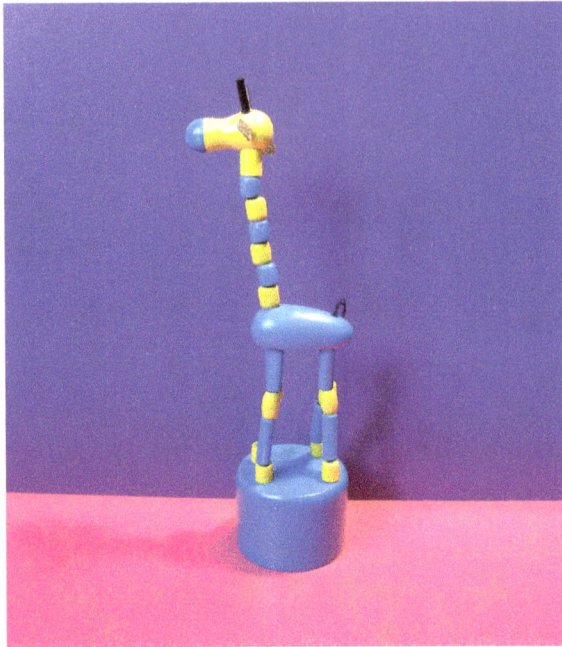

**Figure 1.1**

collapse, when one (imperceptibly) presses the round base of the platform. When one releases the spring again, the giraffe "revives" itself.

From the perspective of physics, the toy is interesting because it can be used to describe *statics*, the first of the three main sections of mechanics — the field of physics where the general laws of motion (*kinematics*) and interaction (*dynamics*) of material bodies are studied.

When studying statics, one considers how various forces act on stationary solids, the conditions necessary to maintain such immobility, as well as the conditions of equilibrium in the field of gravity. The other two branches of mechanics are kinematics, which studies the motion of material bodies without considering the forces acting on them, and dynamics, the main and most general branch of mechanics, which studies exactly how forces cause material bodies to move.

All three sections of mechanics form the basis of *theoretical mechanics* — the basic course of physics, which begins the study of this science in higher scientific institutions. Statics is studied in more detail in higher education, such as in the study of *construction mechanics*, which is already a special subject that is mandatory for future specialists in the field of construction. Here it is worth turning again to our toy, which in its apparent simplicity hides the principle of operation of the main element of many modern buildings.

Construction is one of the oldest areas of human activity, along with hunting, gathering, and farming. People have always required shelter from wild animals, bad weather, or hostile neighbours. The first numerous artificial residential structures of people, which we have had the fortune to study extensively, are the buildings of the Early Stone Age–Paleolithic, whose origins are anywhere from 50,000–24,000 BC. These structures were of the type of huts or canopies built over deepened and leveled soil on flat terrain. In mountainous places, people still lived most often in natural caves, the walls of which were often decorated with surprisingly elegant rock paintings, as evidenced by numerous finds throughout the world.

In the Middle Stone Age (the Mesolithic era), stone buildings were already appearing, the oldest of which is the Göbekli-Tepe temple complex, discovered by nine archaeologists during excavations in modern Turkey, near the border with Syria, the age of which is estimated at about 12,000 years. This structure, along with numerous burial structures of the Late Stone Age–Neolithic, and the Bronze Age, are now known as *megalithic buildings*. At present, they invariably cause surprise because megaliths were structures made of very large stone blocks, the weight of which reached could exceed 50 tons. All megalithic buildings were built of stone blocks, not fastened together in any way, which were a single whole only due to their own weight.

It should be noted that the history of the wet mixture of lime and clay that hardens when it dries is about six millennia old. The earliest concrete resemblant material found by archaeologists, dating back to about 5600 BCE, was used in the construction of a floor about 25 cm thick in a hut in the Danube village of Lapenski Vir, in what is now in Serbia, The Roman architect Mark Vitruvius Pollin (the exact dates of life and death have not been preserved) left us a description of Pozzolan cement and gave recommendations on the use of Pozzolan cement hardening under water in the construction of concrete piers and sea dams.

Ordinary concrete is a mixture of a binder made of sand and stone — nowadays Portland cement is usually used as a binder. In ancient Rome, concrete was a mixture of lime and powdered pumice, ceramics, or Pozzolan. Many structures made of ancient Roman concrete have survived to this day, such as the arched dome of the famous Pantheon, a temple dedicated to the ancient Roman gods restored by order of Emperor Trajan in 126 CE after a fire occurred in 110 CE, this fire being the second after another in 80 CE.

Today, it is not known for certain what methods of calculation were employed by ancient architects in the construction of such complex structures. Written sources of that time have been preserved with project documentation, which provides accurate calculations of the construction time of a particular object, the necessary materials, labour costs, and money. Among some scientists, there is a belief that the calculations of the strength of ancient domed ceilings, spans of arched bridges, aqueducts, and all other elements of buildings and structures were not carried out at all. It is believed that all the buildings of Ancient Greece and Rome were built solely due to intuition, on the practical basis of the method of trial and error — of course, not only their own, but those of many previous generations, counting amongst them the architects of ancient Egypt, Assyria, Babylon, and earlier civilisations. Yet the existence of a certain set of rules and mathematical calculations of the geometry of structures, even without calculating the stress states of the material of buildings, almost certainly existed even then.

The founder of the modern theoretical basis of construction mechanics today is considered to be Galileo Galilei (1564 CE–1642 CE). He studied the changes in the strength of sea vessels with an increase in their size and laid the foundations of a new science in his time on the resistance of materials to acting loads. At the end of the nineteenth century, the main problem of construction mechanics was solved — builders learnt to accurately calculate the internal forces in any element of a building structure and their possible relative displacements under load. This made it possible to widely expand the construction of residential and industrial buildings, roads and highways, bridges, dams, and other complex engineering structures.

The rapid growth of the construction industry was also greatly facilitated by a new construction material: reinforced concrete. This material is an ordinary concrete reinforced with iron bars and wire. Ordinary concrete does not resist bending loads well and works well on compression, so it is usually used in foundations, walls, columns, and similar building structures. Well-stretched iron allows concrete to work on bending, this makes it possible to make all kinds of trusses, floors, beams, and consoles from it.

The widespread and large-scale use of reinforced concrete at the beginning of the twentieth century instigated another improvement for already reinforced concrete structures in order to not only make them hold bending loads, but to also work on tension. This problem was solved by an original invention — in the manufacture of reinforced concrete structures, steel strings or reinforcing rods are placed in a mould for pouring liquid concrete mixture, stretched by a special device, and only then the concrete mixture is laid. After the concrete solidifies, the pre-tensioning force of the reinforcement is transferred to the surrounding concrete, and it is compressed. This creation of a pre-compressed structure makes it possible to significantly reduce the tensile forces on the concrete due to its stretching during operation. Today, prestressed reinforced concrete is widely used in the construction of bridges, high-rise buildings, protective buildings of nuclear reactors, and buildings in areas with a high risk of earthquakes.

Returning to the toy giraffe, one can easily see that its elongated silhouette has a considerable external resemblance to the Ostankinskaja tele-tower in Moscow. More importantly for us, the giraffe has stability due to the same design as the TV tower. Just as the beads of its legs and neck are in a compressed state due to the threads stretched through them by a spring, ensuring the stability of the giraffe, so in the tower, its reinforced concrete shell is compressed by steel cables stretched by special mechanisms (with a total of 149 pieces). The cables are symmetrically located inside the tower, 50 mm from the reinforced concrete body, and provide a total force of more than ten thousand tons. This design of the TV tower, with a height of 540 m, and a base diameter of only 60.6 m is designed for stability even in an earthquake of eight points on the Richter scale and hurricane winds at a speed of 44 m/s.

## 1.2 "Tumbler" (Pertaining to *Balance*)

This toy was most likely invented in Japan. The exact time of its creation is not known, but it has long been called a *Daruma* there. A similar toy in Russia has been known for more than two hundred years, one which we call a "Tumbler". All such toys are arranged according to the same principle — they have a nearly spherical base and a centre of gravity that is below the centre of curvature of the base. Such a curvature ensures the return of the toy to a stable equilibrium position after any attempt to deflect the toy at any angle and in any direction.

With the help of this toy in theoretical mechanics, one can find key insights into the section of statics concerning "equilibrium and stability of solids". The definition of equilibrium conditions is described as when the equilibrium of bodies is preserved by two conditions being met: the vector

**Figure 1.2**

sum of all external forces is equal to zero, including the moments of forces (the product of the force value by the distance from the point of its application to the point of support) acting on the body under consideration. Mathematically, this condition is written as:

$$F_1 + F_2 + \cdots + F_n = 0 \quad \text{and} \quad M_1 + M_2 + \cdots + M_n = 0.$$

There is also a different, somewhat simplified rule for the equilibrium conditions of a solid body that has support from a surface. According to this rule, a solid body is always in a static equilibrium position, provided that the vertical line passing through its centre of gravity is within the area of support. This rule helps demarcate three types of equilibrium: stable, unstable, and indifferent.

A solid body will always maintain a *stable balance* if the vertical line lowered from the centre of gravity passes through the support area and does not coincide with the line bounding this area. The most obvious example of such an equilibrium is a cube lying on a table — its center of gravity is always above the area of support.

The body will always maintain an *unstable balance* if the vertical line lowered from the centre of gravity of the body coincides with the line limiting the area of the support. Acrobats at a circus, for the purposes of this example the representatives shall be tightrope walkers, amaze us with their ability to move on a tightrope with a minimum support area, demonstrating unstable balance as clearly as possible.

The body will always maintain an *indifferent balance* if a vertical line passes through its centre of gravity at any change in the position of the body in space. A typical example of an indifferent balance is an ordinary wheel or ball when they are moving on a flat surface — they can be stopped at any time, and they will always be in balance.

# 1.3 "Shooting Pneumatic Toys" (Pertaining to *Kinematics*)

Toys, such as those shown in Figure 1.3, can shoot small plastic pellets, propelling them with a jet of compressed air. The air pressure is provided by a pre-compressed handspring. The small mass of the projectile, even at a reasonably high speed, for example of 40 m/s, provides very little energy, less than 3 J, which makes them innocuous. Nonetheless, recklessness with projectiles is discouraged by the manufacturers of the toy, and supervision with children is recommended. The pellet in flight moves too quickly to be observed but makes a perfect example for discussing the speed of moving objects.

The word "speed" is intuitively associated with the movement of something in space, but this word is often used much more widely. For example, in computing, the expression "data transfer speed" is used despite the rate of the transmission of data not involving the relocation of data. Similarly with the expressions "speed of thought" or "breakneck speed", this idea of speed is figurative rather than actual, nothing is in motion. The concept of speed for a layman is often associated with punctuality and brevity as well as with motion.

**Figure 1.3**

As a physical concept associated with the change in the position of a material body in space, the contemporary definition of the term "speed" is relatively young. A long time ago, among some philosophers the definition of the concept of "movement" caused fierce discussion. There have even been philosophical trends that denied the possibility of movement. Despite the absurdity of this trend, today the paradox of the tortoise and Achilles is widely known by the name of author Zeno of Elea (fifth century BCE). In this paradox, it is stated that even if Achilles were to run 100 faster than the speed of the turtle, due to his being initially behind it by 100 m, he will not be able to catch up with the turtle crawling away from him. The explanation is that while Achilles runs 100 m, the turtle will crawl away from him by 1 m, which will be the new distance between them. If Achilles moves forward another metre, the turtle will also crawl forward one centimetre, and again there will be a very small distance between them. No matter how fast Achilles may close the previous gap, the gap will always grow. The distance between them will always be reduced, but you cannot catch up with the crawling turtle, because there is no limit to the division of units of time and distance. In practice, the paradox does not justify itself, since the categories of time and space are discrete.

The mathematical notation of speed, as distance divided by time, was not yet accepted in ancient times as it is today. In those times, when estimating the speed of movement, one would only compare either the distances travelled in the same time intervals, or vice versa — the times required to overcome the same distance. Aristotle defines uniform motion as the motion in which "bodies travel the same distances at equal times".

Isaac Newton (1642 CE–1727 CE) in the mid-1660's — and independently, Leibniz in 1675 CE — developed the mathematical concepts and techniques for differential and integral calculus, giving a modern interpretation of the concept of instantaneous velocity as the distance travelled per unit of time. In rectilinear motion, the nature of speed change may be different. Velocity ($V$) can be constant — with $V$ being constant, such motion is called uniform, and the distance traversed by the body during such motion is determined by the formula $S = Vt$. In the generally accepted international system of units, the units of speed measurement are metres per second (m/s). For various technologies, other methods are also widely used, for example, the use of kilometres per hour (km/h) and miles per hour (m/h) in the automotive industry.

Speed can increase or decrease over time as well, which is described as *acceleration*; typical notation is *a* or *W*. Expressed mathematically, when the speed change is equal to $\Delta V$ over a period of time *t*, the average acceleration $a_{sr}$ is determined by the formula: $a_{sr} = \frac{\Delta V}{t}$. The units of acceleration in the international system of units are meters per second squared (m/s$^2$). With a constant increase in speed, $a = const$, the movement is called *equidistant*. If the acceleration *a* is positive (thus marked +), the speed is written as $V = V_o + at$, where *V* is the current speed, $V_o$ is the initial speed, *a* is the acceleration, and *t* is the time since the beginning of the accelerated movement. The distance *S* traversed by the body in this motion is determined by the formula: $S = V_o t + \frac{at^2}{2}$. If the speed is uniformly slowed down, then in mechanics this means that the movement is still accelerating, the difference is only in the sign of the acceleration now being negative (thus marked –). Equidistant movement is described by the formula $V = V_o - at$, where *V* is the current speed, $V_o$ is the initial speed, *a* is the acceleration, and *t* is the time since the beginning of the deceleration of the movement. The distance *S* traversed by the body in this motion is determined by the formula: $S = V_o t - \frac{at^2}{2}$.

Unfortunately for ease of calculation, movement is rarely rectilinear. Almost all material bodies in our life move along curved trajectories. Even if we are moving along an arrow-straight highway, we are moving along a curved trajectory with a huge radius of curvature equal to the radius of the Earth. The simplest case of curvilinear motion is rotation. In theoretical mechanics, the kinematic characteristics of rotational motion are the direction of rotation, angular velocity, and angular acceleration. The average angular velocity $\omega_{sr}$ is the ratio of the angle of rotation $\Delta\varphi$ to the time interval $\Delta t$ for which the turn occurs, written as: $\omega_{sr} = \frac{\Delta\varphi}{\Delta t}$. The units of angular velocity in the international system of units are the inverse seconds (s$^{-1}$). The absence of a unit of measurement in the numerator is explained by the fact that the angles of rotation in the international system of units are measured in radians, which are not a dimensionless quantity, thus the ratio of radians to angular degrees is determined by the formula:

$$A(\text{radian}) = B(\text{degree})\left(\frac{\pi}{180}\right)^o.$$

The instantaneous angular velocity $\omega$ is the limit of the ratio of a very small rotation $\Delta\varphi$ for an equally small period of time $\Delta t$ when the time interval tends to zero $\omega = \lim_{\Delta t \to 0} \frac{\Delta\varphi}{\Delta t} = \frac{d\varphi}{dt}$. Angular acceleration $\varepsilon$ is similar to linear acceleration and characterises the change in angular velocity $\omega$ per unit of time. Its sign is positive when the angular velocity increases and negative when it decreases. The numerical value of the average angular acceleration is determined by the formula: $\varepsilon_{sr} = \frac{\Delta\omega}{\Delta t}$; the instantaneous angular acceleration is determined by the formula $\varepsilon = \lim_{\Delta t \to 0} \frac{\Delta\omega}{\Delta t} = \frac{d\omega}{dt}$. The units of angular acceleration are the inverse of seconds squared.

Rotation with a constant angular velocity, $\omega = const$, is characterised by the fact that the time of one complete revolution, the so-called rotation period $T$, is a constant value. Since one complete revolution is a 360-degree rotation, or $2\pi$ in radians, the angular velocity of a uniform rotation can be written as $\omega = \frac{2\pi}{T}$.

The number of revolutions per unit of time is also called the rotation frequency $v$ and this rotation characteristic is inversely proportional to the time of one revolution $v = \frac{1}{T}$. The angular velocity determined by the rotational speed is written $\omega = 2\pi v$.

In engineering, angular velocity is often measured by the number of revolutions per unit of time and denoted as $n$. The measurement formula in this case will take the form $\omega = n2\pi$ and the units of measurement in this case are revolutions per second (r/s).

Any point of a rotating body has a linear velocity $V_t$ directed tangentially to its circular trajectory. Such a speed in mechanics is called *circumferential* and has a direct relationship with the angular velocity $\omega$, which is expressed by the formula $V_t = \omega R$, where $R$ is the radius of circular rotation. It is this relationship that allows one to measure the high speed of a pellet with a rather ingenious technical solution. The aforementioned toy provides a great example for these rules for predicting motion; finding a solution to the speed of the pellet only requires a young shooter using the pneumatic toy and a bit of number crunching. The accuracy of the measurement in your experience will be very approximate, but the main goal of the lesson from this toy will be an explanation of the measurement principle itself based on the relationship between the linear, circumferential, and angular velocity of rotation.

For the experiment, you will need a ski pole, two discs made of craft paper (with a diameter of about 20 cm), a cord 1.5 m in length approximately, and two chairs. The general scheme of the experimental setup is shown in Figure 1.4.

The craft paper discs are tightly attached to the ski pole perpendicular to its axis and at approximately 1 m apart from each other. About 15 to 20 turns of the cord are wound on the ski pole, and it, together with the craft paper discs, is placed on the back of the chair as shown in Figure 1.4(a). If you pull the cord evenly, the ski pole will unwind, which will cause it to rotate together with the craft paper discs with a constant angular velocity $\omega$. During the unwinding of the cord from the ski pole, approximately one to two seconds, it is necessary to count the number of windings of the cord. Dividing the number of the number of coiled turns by the time of coiling, you can determine the number of revolutions of the rotating discs $n$ and then the angular velocity $\omega$ can be easily calculated by the formula: $\omega = n2\pi$.

As an example, with three windings in one second, the angular velocity will be equal to $\omega = 3 \times 2 \times 3.14 = 18.84$ s$^{-1}$. To measure the flight speed of the pellet $V_1$, it is necessary to shoot using the pneumatic toy at the cardboard discs while they rotate so that the plastic pellet breaks through both disks.

Remember, it is important to aim carefully, ensuring the most accurate flight path of the pellet with the line parallel to the axis of the ski stick.

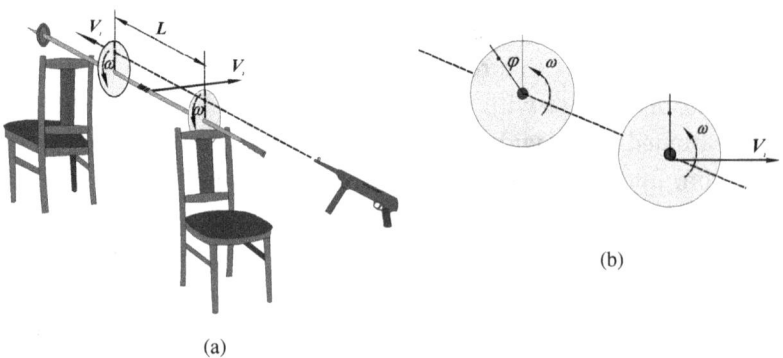

(a)

(b)

**Figure 1.4**

The more that the trajectory of the pallet is parallel with the ski pole, the more accurate that this measure will be. The pellet, having broken through the first disc, will have to fly a distance $L$ after which it will break through the second disc. The time it takes for the pellet to fly along the ski pole between the discs is very short, and it is very difficult to measure it, but this is not necessary as it can be calculated.

While the bullet, having punched a hole in the first disc, covers the distance $L$, the ski stick, with the discs rotated at some angle $\varphi$, will be recorded by the hole, pierced by a pellet on the second disc, as shown in Figure 1.4(b). Knowing the angular velocity of rotation and the angle of rotation $\varphi$ in radians, one can use the formula $\omega_{sr} = \frac{\Delta\varphi}{\Delta t}$ and, assuming that $\omega = \omega_{sr}$ and $\varphi = \Delta\varphi$, one can easily find the numerical value of turn time $\Delta t$.

As an example, suppose the angle of rotation, measured in degrees, was equal to $\varphi = 30°$ degrees. To translate it into radians, one must use the formula: A (radian) = B (degree) $\left(\frac{\pi}{180}\right)^0$, thus $\varphi = 30\left(\frac{3.14}{180}\right) = 0.52$ radians. The turn time in this case will be $\Delta t = \frac{\Delta\varphi}{\omega_{sr}} = \frac{0.52}{18.84} = 0.028$.

Knowing the time for which the pellet covers the distance "$L$", it is easy to calculate the speed of the pellet using the formula $Vsr = \frac{S}{t}$, assuming "$L = S$" and "$t = \Delta t$", we find that the speed of the pellet fired from the pneumatic toy in our hypothetical case to be $V = \frac{L}{\Delta t} = \frac{1}{0.028} = 35.71$ (m/c) or 9.92 km/h.

## 1.4 "Pecking Woodpeckers" (Pertaining to *Friction*)

This toy is made from a vertical steel rod along which a wooden washer may balance itself by a skewing of the bore of the washer to whichever side a wooden woodpecker is suspended by a spring off of said washer.

   Whilst stationary, the friction between the rod and washer is sufficient to suspend the woodpecker, thus the woodpeckers and their washers appear to be floating around the rod. Once set into motion, the woodpecker will work their way down the rod under the influence of gravity. The motion of the washer down the rod stops and starts as it descends. This is due to the skew of the washers briefly approaching being parallel during the oscillation of the woodpecker on its spring, reducing contact between the washer and the rod beyond the point of sufficient friction to suspend the woodpecker's apparatus, but soon after this approximate

Figure 1.5

alignment the woodpecker will rebound and re-skew their washer, which will briefly steady the washer until the oscillation of the woodpecker returns it to approximate alignment, which will happen due to the inertia of the woodpecker and the transmission of this force by the spring to the washer. Nonetheless, this does not explain why the contacting surface creates a force such as friction.

Friction is of an electromagnetic nature. When the surfaces of different bodies become as close to each other as their fields will allow, the electromagnetic fields of the bodies' molecules (most significantly the superficial ones) interact with each other. These interactions will vary from body to body depending on their constitution.

There are three types of friction:

- *Static friction*: when shear force is unable to overcome the attraction between bodies,
- *Sliding friction*: when shear force is able to overcome the attraction between bodies,
- *Rolling friction*: when bodies roll over each other's surfaces (examples being a line of contact between cylinders or a point of contact between spheres).

In Da Vinci's manuscripts dating back to 1493, one can find records that they noticed the dependence of the friction force on the force of pressing the contacting surfaces together rather than on contact area. Centuries later, Guillaume Amontons in 1699 and Charles Coulomb in 1785 established the dependence of friction on external parameters, which is now knows as the Amontons–Coulomb law. According to this law, friction force is defined by: $F_{tr} = fN$, i.e., the product of the coefficient of friction $f$, determined by the roughness of the contacting surfaces, by the pressure force $N$, perpendicular to the contact area. As Da Vinci had pointed out hundreds of years prior, the contact surface area does not influence the value of the friction force.

Friction is one of the most common forces which we encounter in our lives. It is often useful, allowing us to hold things and to balance. In many technologies, we attempt to mitigate friction to require less energy to maintain motion and to reduce wear.

By increasing the roughness of contacting surfaces, one may increase friction. Sometimes, for the purpose of increasing friction, surfaces may be smoothed rather than roughened such that the contacting surfaces can increase their electromagnetic interaction. One can reduce friction by using various lubricants, replacing sliding friction in mechanism with rolling friction, as well as using materials for contacting surfaces which have an inherently low coefficient of friction. Some examples of substances with low coefficients of friction would be cast iron on cast iron ($f = 0.12$), teflon on steel ($f = 0.05$), and steel on ice ($f = 0.015$). Nowadays, ceramics based on an alloy of boron, aluminium, and magnesium are considered the most slippery surfaces, outdoing even teflon ($f = 0.02$).

# 1.5  "Toy Scales" (Pertaining to *Weight and Mass*)

This set of toy scales is not an instrument suitable for any serious purposes, however they will allow one to weigh small objects with adequate
accuracy for the purposes of demonstration. This toy demonstrates the
rules of the lever. As seen Figure 1.6, one unit of weight on the left is balanced against two units of weight on the right, which is explained by the
distance between the pivot being twice as far for the single unit as for the
two units. The rules of the lever will be further explored in the description
of the toy "Propeller on a Screw".

Scales are often incorrectly conveyed as determining the weight of a
body. Scales actually measure relative mass. Scales will work by the same
logic under different conditions of gravity: for the same bodies in the same
relation to each other on the lever, they will show the same relative mass
despite the bodies having different weight than what they would under
different moments of gravity.

Earlier in human history, determination of weight was one of the most
crucial issues for trade, seen as fair trade required accurate assessment of

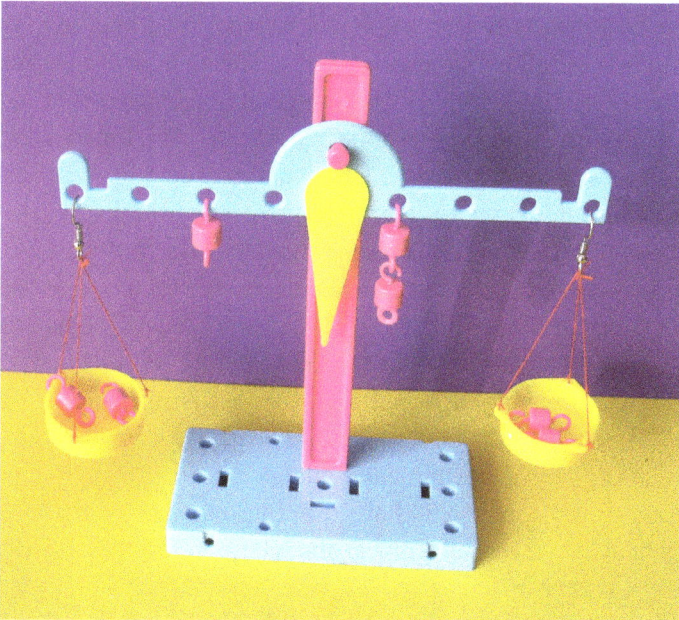

**Figure 1.6**

quantity. The first scales, taking a form similar to those of our toy with two bowls equidistant apart from their pivot, appeared two thousand five hundred years ago in ancient Mesopotamia and Egypt. Half a millennium later, in India and China, the people there learnt to measure weigh with the help of unequal shouldered scales. The body would be suspended off a hook which was on the end of a heavy metal rod; the rod was suspended through a belt loop. The weight of the load would be determined by moving the loop along the rod until the rod reached a horizontal position. The suspension point of the belt loop indicated the weight of the load according to the markings previously applied to the rod. In this case, the belt loop acts as the pivot, the shoulder itself constitutes the relative mass.

In ancient times, the weighing technique was already sufficiently developed for the concerns of contemporary livelihood, thus scientists of this time were more concerned with the philosophical questions of what matter is, what attracts matter to the earth, and what makes it weighty. Aristotle observed that two objects of different weights when dropped from the same height would have the heavier object fall first. He explained this as each particle of mass accumulating its drive to reach its natural level of the Earth. Despite Aristotle being indisputably one of the greatest scientific minds of humanity, his theory was not infallible. In the case of his description of falling, he was incorrect. The authority of Aristotle and the nature of intuition left this misunderstanding unmoved. This was until Galileo Galilei questioned Aristotle's assertion and found it wanting. Galileo's famous experiment was a revolutionary step forward in the scientific method, attempting to control as many factors as possible to isolate the factor upon which was being experimented. He had two balls of the same dimensions and different masses dropped from the Leaning Tower of Pisa and found those balls to fall at the same rate. Galileo explained Aristotle's results by air resistance. This has often been demonstrated in schools by the dropping of a lead pellet and a Styrofoam bead (or other such bodies) in a container out of which the air has been pumped.

Galileo Galilei was instrumental in understanding why falling bodies of differing dimensions would have differing rates of falling but left the question of why they would fall at the same rate if they had identical dimensions. Isaac Newton came to answer this question with their formulation of the law of universal attraction. They suggested that weight appears in any body only when it is suspended, such as if an object's weight rests in one's hand as one holds it, and that when an object has no resting force then it will be in free fall, in which case it will be weightless. The difference between weight and mass is fundamental for understanding these forces at rest and in free fall. *Weight* as a force comes from the gravitational attraction between bodies (the dominant body of attraction is the Earth; however, all bodies exert attraction on each other); mass is a characteristic of bodies themselves. Newton considered mass as the amount of matter within a given body's volume. Nowadays, the concept of mass has changed: mass is a measure of the inertia of a substance, which is to say the ability to resist changes to the vector of acceleration, as well as the ability to be subject to gravity.

It is still not understood how bodies attract each other; the nature of gravity remains a mystery of great importance. Newton described the laws of gravitational attraction within the framework of classical mechanics, which we shall explore. Let us consider how these patterns manifest themselves in the example of the interaction of two bodies — let one of them be the Earth, its mass is equal to $M$ and the other an arbitrarily selected body with mass $m$. According to the law of universal gravitation between two bodies of masses $M$ and $m$, the gravitational attraction force $F$ is equal to $F = G\frac{Mm}{R^2}$. $G$ is the gravitational constant, which is determined experimentally, is equal to $6.67 \times 10^{-11}$ m$^3$/s$^2$ kg and is necessary to bring the units of measurement into compliance, and $R$ is the distance between the masses, which can be considered equal to the radius of the Earth (which is 6.378.137 m). When the bodies under consideration approach under the influence of the gravitational attraction force $F$, and this will correspond to the fall of the body $m$ to the Earth, and the movement of the Earth to the body $m$, the acceleration of each of them can be calculated by dividing the interaction force $F$ by the corresponding mass. For the acceleration $w$ of the body $m$, we get the formula $w = G\frac{M}{R^2}$, the acceleration of the Earth will be equal to $W = G\frac{m}{R^2}$. Comparing these two accelerations, we can draw the following conclusions:

- The height of the fall of the body $m$ is far smaller than the radius of the Earth, thus the distance $R$ can be considered effectively unchanged,
- The mass of the Earth $M$ is far greater than the mass $m$ of the falling body, thus the acceleration $W$ of the Earth in comparison with the acceleration of the falling body can be considered negligible,

- The acceleration of any falling body $w$ can be considered constant, since the mass of this body is not included in the formula that determines its value. $G$, $R$ and $M$ are constant values, and $W$ is negligible.

Thus, Newton proved that all bodies fall to our Earth with the same acceleration no matter their mass (for extremely small bodies, which, compared to our Earth, all bodies of significance to present human endeavours are). The exact value of acceleration of gravity, $g$, differs between locations around the world due to various factors such as centrifugal force from our Earth's rotation, the deviation in the form of our Earth from a sphere, and tidal influences from the attraction to our Moon. The average value of acceleration for the gravity of our Earth is assumed to be $g = 9.80665$ m/s$^2$.

Assuming the value of $g$ and knowing the value of $m$ for a given body, one can determine weight $P$ as $P = mg$. The mass of a body under the international system of units is typically measured in kilograms, but seen as weight is a force, it would not be incorrect to alternatively measure this in Newtons. However, for all scales, the measuring scales are marked in such a way that they show not the weight, but the mass at once, i.e., the quotient of dividing the weight by the acceleration of gravity. Since it is assumed that one Newton is equal to 98.0655 g, then one kilogram of mass shown on any scale corresponds to a weight of 9.80665 N.

The kilogram as a unit of measurement is interesting because it is the only unit based on a particular object being the standard for its mass. This object is a platinum–iridium alloy cylinder which is stored in the Paris Chamber of Weights and Measurements. Other units of measurement are determined from the fundamental laws of nature, such as temperature (degrees Kelvin, one degree is one hundredth of the calories to heat water from ice to steam).

To conclude, consider cosmonauts on the International Space Station. Their mass is not diminished by their being in orbit, and they never cease being acted upon by gravity from our Earth, however they cannot be weighed as their centrifugal force has a vector which almost perfectly counters the vector of gravity. With an orbit altitude of about 380 km and a radius of our Earth of 6,371.2 km, the difference in the gravitational attraction force on our Earth's surface and in orbit is determined as 0.00000000269520 N. This force is minute to the point of being negligible, but this reduction in weight is totally separate to the question of mass.

# 1.6 "Propeller on a Screw" (Pertaining to *Simple Mechanism*)

This toy can be very entertaining despite its simplicity. A light propeller can be launched into the air with the help of a slider which can run the length of a spiralling plastic strip similar. This is one type of simple mechanism.

The most ancient simple mechanism may be the lever, upon whose principles the gate and the pulley were invented. These mechanisms are so ancient that it is not always clear which came first and who conceived of them. Archimedes is considered to be the greatest mechanic of the ancient world; they may have invented the screw. They did invent a mechanism for lifting water which involved an inclined plane rolled into a spiral, which may have been the first time such a mechanism was used.

Heron of Alexandria wrote in their text *Metrica* on these things which they termed simple mechanisms, investigating them individually and combined. Even in their work (written during the first century CE), they remarked on the importance of the moment of force and recommended that friction be considered when working with complex mechanisms.

All simple mechanisms share a principle of operation: with the same effort by a person in the process of an action, simple mechanisms can

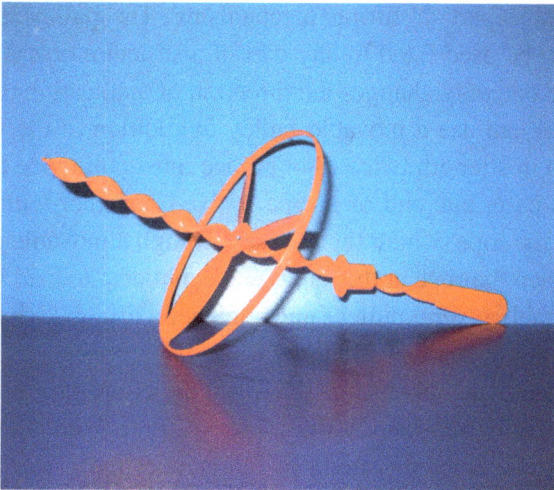

**Figure 1.7**

change the direction and magnitude of the vector of force enacted upon a body. This is based upon the golden rule of mechanics: the gain in strength is proportional to the loss in distance. The lever is a clear example of the logic of this golden rule. The distance from the support point of the lever and the lifted load should be much less than the distance from the same support point to the point of application of force. This ratio of distances allows, by applying a small force and moving the lever over a considerable distance, to obtain a much greater force at the other end of the lever, although moving the load much less. One example of a lever can be considered an ordinary shovel. By moving the handle with a small force at a distance of about 50 cm, you can apply a much greater force to the ground layer, however the movement will be proportionally far less, being no more than 10 cm.

The logic of a gate is demonstrated in the way which one lifts water in a bucket on a rope from traditional wells. A cylindrical piece of a log with a rope wound around it has a radius much smaller than the length of the handle for which it rotates. This difference in the length of the rotation of the handle against the length of the rope withdrawn as it is stretched by the weight of a bucket of water allows one to reduce their effort over time. It is possible to rotate the gate with relatively little effort, even on a relatively long length of the rope. According to the same principle, pulleys can be arranged which, together with cables or chains thrown over them, are an indispensable part of lifting mechanisms. The pulley is interesting because it can be used fixed to any part of a structure or natural support, in which case one only changes the direction of application of the pulling force. But one can use a movable pulley in addition to the fixed pulley, getting a gain in strength twice (though one must pull twice the length of rope). In this case, one end of a rope is fixed, together with a load suspended from the rope, which then passes through a movable pulley at the other end. When the rope at this other end is pulled, the load is lifted. You can also assemble several pulleys and engineer a lifting device that allows one to get multiple gains in strength.

An inclined plane allows one to reduce the effort over time of elevating an objecting by reorienting the vector of force from vertical towards horizontal. The smaller the angle of inclination, the greater the gain in force, but the greater the distance you will have to move the load horizontally (which will in fact increase overall effort but reduce the threshold of effort at any given time). A significant disadvantage of the inclined plane is a large friction force, so usually additional rollers or lubricant are used. A famous example of this comes from the construction of the Egyptian pyramids: clay mortar was poured under stone blocks to reduce the friction force. The wedge is used when it is necessary to divide strong but brittle objects into parts. Another example that can be seen in the Egyptian pyramids is how the sandstone was hewn. The spiral, in addition to the traditional use when moving or lifting liquid and bulk substances, has found the widest scope of application in the manufacture of a variety of fasteners, such as nuts, bolts, screws, and the like.

# 1.7 "Wind-Up Toy Car" (Pertaining to *the Wheel*)

The wheel could be considered as the progenitor of simple mechanisms, and the history of its invention is lost in the mists of time. The wheel was most likely invented by rolling heavy loads with the help of a log, and as processes for refining materials were themselves refined, the wheel as a machine was created. With the wheel came the advent of transport, giving rise to the possibility of far more efficient material distribution.

The wheel's first limitation was humanity itself. Unable to muster great strength compared to their ambitions, early people turned to domestication of animals not only for the purposes of provision of food but also for use in labour. Many beasts were out to the task of pulling heavy loads, be they ox or ass or elephant, but none were so famous as a beast of burden as the horse. The first horse drawn cart appeared about four thousand years ago in Mesopotamia, resembling chariots with their single axel. In Ancient Egypt at the time of the invasion of the lower Nile, horse-drawn carriages were depicted to have been used already.

**Figure 1.8**

The first wheel mechanism that could move independently and for quite a long time was invented by Leonardo Da Vinci. Instead of requiring a draft animal, an elastic spring was used as an energy source. Nowadays it is not known for sure whether the idea was implemented in practice by Da Vinci themself, however, in the modern age, successful reproductions from their schematics have been achieved. The most successful attempt to reconstruct the spring carriage of Leonardo Da Vinci turned out to be a working mechanism assembled in 2004 CE, in which the Italian scientist Carlo Pedretti used powerful spiral springs to move, and two more flat springs automatically controlled the direction of movement. Unfortunately, the efficiency of the spiral spring is not great — even with effective lubricants, it is unable to store and release more than 70% of the energy put into it for the work it can perform. Furthermore, the reserve of elastic energy that the best spring can store in combination with its own weight does not make it possible to use for anything of great scale and functionality, though it suffices for the purposes of toys.

In the seventeenth century CE, the automatic running mechanical cart was created in Russia by a self-taught peasant by the name of Shamshurjenkov Ljeontij Luk'janovich. They designed this cart in 1752 CE, constructing it with four wheels to be pedalled by two drivers and to carry two passengers. The cart could reach a momentum of 15 km/h (as was indicated by its speedometer, also invented by Shamshurjenkov). Kulibin Ivan Pjetrovich further refined this instrument in 1791 CE. Their three-wheeled cart only required one driver and contained a flywheel that could store energy when descending for when one would be ascending. The average speed of this cart was 10 km/h, which for its time was quite impressive. Even today, spring-powered batteries have promising applications. The next stage in the development of pedalled carriages was the invention of two-wheeled bicycle.

# 1.8 "Toy Bike" (Pertaining to *Green Transport Technologies*)

This toy is also known as a finger bike. It is constituted from a metal frame with replaceable rubber wheels on bearings with an easily rotated steering wheel. Children can play with these and perform tricks with the instrument using their hands. The logic of its mechanism is similar to BMX bikes; they are good for developing dexterity and coordination.

The invention of the bicycle is sometimes attributed to Leonardo Da Vinci; however, this has been proven to be an incorrect attribution in a number of historical studies, famously in David Herlihy's *The History of the Bicycle*, amongst others. Nowadays it is believed that the first patented invention of a two-wheeled, human-powered vehicle was the trolley designed by the German baron Professor Karl von Drez in 1817 CE. This trolley was a wooden scooter with two longitudinally arranged wheels. The rider, sitting on the frame, could move their legs and alternately push

**Figure 1.9**

off from the ground. It allowed the rider to move twice as fast as a human running without requiring nearly the same effort; the gyroscopic effect of the wheels' rotation was able to stabilise the rider and their vehicle. Scooters such as this trolley became very popular, and their industrial production was established.

By 1840 CE, numerous improvements had been made which improved comfort and convenience, such as the inclusion of pedals and a saddle, however this still did not resemble the modern bicycle. A little later, the Frenchman Pierre Lalman simplified the design, transferring the pedals to a large driving front wheel. The success of this design led to the mass production of two-wheeled carriages in 1864 CE and to the name "bicycle" entering the vernacular language. Within a few years the modern structure of the bicycle emerged. By 1867 CE a metal wheel with spokes appeared; by 1878 CE the chain transmission replaced the metal rod and hinge systems of the 1840's. The bicycle as we know it was patented in 1886 CE by the Englishman John Starley.

Learning to ride a bike, while euphemistically unforgettable, can be a difficult task. Since there are only two points of support during operation and one's centre of gravity is not on numerous axes, one must build sufficient momentum to cause a gyroscopic effect. When mounting a bicycle, one rarely is at rest, usually one shall propel oneself forward a little to help rapidly reach equilibrium. An experienced cyclist may start moving from being in the saddle at rest, however they will be in a state of unstable equilibrium in those beginning moments (as is described in the description of the toy "Tumbler").

The improvements in material refinement technologies in the last few decades have allowed humanity to create ever more efficient, light, and sturdy bicycles. Carbon fibre and titanium alloys have allowed improvements to weight without sacrificing solidity; microprocessors and batteries have allowed the reservation of more of the energy that is expended to be used when demands increase, much like the functionality of Kulibin's flywheel.

Nowadays, many enthusiastic designers are attempting to re-imagine these human powered vehicles, making simple, eco-friendly, safe transport technologies. However, the design of the bicycle as patented by Starley has not seen radical reimagination in a long time; why reinvent the wheel?

# 1.9 "Gravity Cart" (Pertaining to *Perpetual Motion*)

This toy, invented about 2,000 years ago, is not at all difficult to make yourself. Two pairs of wheels on two axles which can rotate freely in sliding bearings are fixed in the base of a light truck. A shoelace is tied to one of the axles fixedly and led through a pulley, which is raised to a specific height above the base of the cart. A small weight is tied to the other end of the lace. The length of the lace should be such that when it is fully wound, the axis of the weight is raised to the maximum height. When the cord is completely wound off the axis, the weight drops almost to touch the base of the cart and hangs on the stretched cord. This is the entire mechanism; it is very simple. The toy can be set to motion on a flat surface. To begin motion, completely wind the cord on its axis and then leave the toy to itself. The weight, sinking under its own weight and unwinding the cord from the axis, will make the wheels rotate and the cart will move as if by itself. The movement will only last as long as the string can unwind, then the cart will stop. But it is not difficult to set it in motion again, it is enough to wind the cord again and release the cart with the

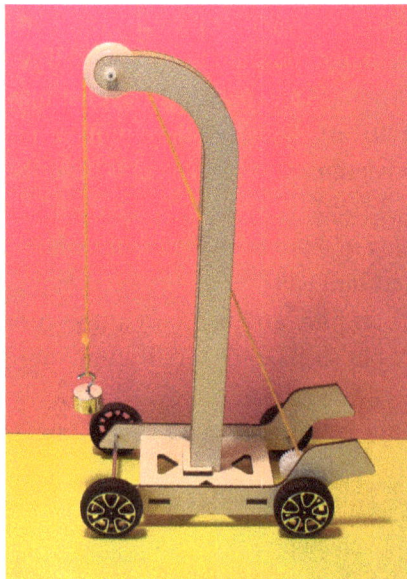

Figure 1.10

raised weight. The toy is interesting because all the energy for the movement of the toy is stored in the potential energy of the suspended weight, rather than in anything so obvious as a combustion engine or a battery. Here it becomes clear how simple mechanisms can transform the unidirectional force of attraction that is gravity into perpendicular motion.

Heron of Alexandria invented such a mechanism as the one which is being investigating. The description of the toy "Heron's Chain" will further elaborate on their ingenious inventions. The main feature of Heron's inventions is their conspicuousness; they wanted to astound and conceal the logic of their mechanisms within their forms. The principle of movement of a toy cart was applied by Heron to create a self-moving pedestal on which ancient divine characters solemnly rode on the stage of ancient theatres. Such a "magical" appearance of the characters in theatrical performances caused special feelings of exaltation among the audience, who then usually watched dramas and plays of a religious orientation. The mechanism that makes the pedestal move, of course, was hidden by Heron from the audience and it was surprisingly simple. A heavy load, connected by ropes and pulleys with wheels, was placed on the surface of a large container with wheat grain. As the grain poured out through the hole in the bottom of the container into another, lower container, the load on the grain descended and pulled the rope wound on the wheel axle, rotated it, and forced the entire structure to move.

To the present day, many inventions submitted to patent offices around the world use gravity as a source of power. There is even a separate class in patent claims under the title F03G pertaining to instruments powered as such. It is to this class where many notorious perpetual motion machines end out being relegated. There notoriousness has done nothing to dissuade inventors from attempting to make them, despite the fact that in 1775 CE, the Paris Academy of Sciences unanimously decided to refuse to consider such projects, defining them as "perpetual motion engines of the first kind". Many great scientists, including but not limited to Leonardo Da Vinci, Robert Boyle, and Johann Bernoulli, have attempted to reach this mythical pinnacle of engineering. The seventeenth and eighteenth centuries were full of inventors attempting to defy entropy; no one's ingenuity was able to surpass this hurdle. There were three types of not quite perpetual motion machines that came into being:

- *Those powered by gravity*, often by the rotation of unbalanced loads, the implementation of the Archimedean screw, the pouring of liquids, etc.
- *Those powered by hydraulics and pneumatics*. Mercury was especially popular here, flowing through intricate pipelines under its own weight. There are known attempts to use the property of capillaries to raise water. We tried to find a way to extract useful work using the hydrostatic paradox, the essence of which is that two vessels of the same volume have different pressure on the wide and narrow bottom. The same level of fluid in communicating vessels of different volumes gave hope to other supporters of the idea of non-stop movement.
- *Those powered by electromagnetism*. Usually, efforts were made here to asymmetrically weaken the magnetic field strength of natural magnets.

While some efforts to make perpetual motion machines may have been sincere, there were certainly numerous cases that were not sincere as well. Where Heron of Alexandria's self-propelled chariot concealed its function seemingly for the purposes of wonder, the hucksters of the enlightenment would find ways to conceal elaborate springs, beasts, or even people within their fake perpetual motion engines for the purposes of garnering wealth. The first emperor of the Russian Empire, Peter the First, was always fascinated by Western technologies and tools and was himself conned by the German Johann Bessler. Its mechanism was a wheel with a complex system of levers and pendulums hidden inside, which, once put to motion and left in a closed and sealed room, would seemingly continue its motions for remarkable extents of time. It turned out to be powered by people.

In 1841, the German naturalist Robert Mayer, for the first time, clearly formulated the law of the conservation of energy. One may have thought that all disputes on the topic of perpetual motion should have been laid to rest, but alas. To this day, the task of refuting one of the fundamental principles of physics has not been relinquished. Considering the apparent realities of our time, the projects of perpetual motion engines have undergone drastic conceptual shifts as the metaphysics of modern sciences develops. For example, there are perpetual motion projects which incorporate magnets, where, unlike in the already antiquated hypotheses, a design is proposed that now certainly should work thanks to a combination of both conventional magnets and those operating in the superconductivity mode. At the same time, the loss of energy necessary to maintain the operation of such an engine should come from the transmission of heat from ordinary magnets and the creation of ultra-low temperatures that provide superconductivity conditions. With the obvious technical impracticability of such a project, it is a "perpetual motion engines of the second kind". In such devices, one type of energy is converted into another without any losses during the transition. The creation of such machines is not possible as it contradicts the second law of thermodynamics. In nature, the processes of independent transfer of thermal energy from less hot bodies to more hot bodies are impossible. If one were to have a practical implementation of such a transition, an external energy supply is necessary, a good example of which is a household refrigerator.

However, something similar to perpetual motion is still possible, in that the energy which powers these engines can be present in the environment in which these engines are situated regardless of these engines' presence. Hydroelectric dams, solar cell relays, eoliennes (wind turbines), and geothermal systems all draw on the forces of nature much as life does too. The sun, the moon and tectonic activity all are possible to harness, though they are not eternal. In human terms, these structures are so vast that they may as well be eternal, they have predated life and life is contingent upon them.

# 1.10 "Heron's Chain" (Pertaining to *the Physics of Antiquity*)

When Heron of Alexandria lived is up for debate, but prevailing opinion is that they lived in the first century CE. What is undebatable is how important their mathematical and mechanical forays were to the development of what would one day become physics and engineering. Their works have not survived in their entirety to the present day. The works which are known are *Mechanica, Catoptrica, En Dioptra, Pneumatica, Belopoeica, Automata*, and *Metrica*. Heron's mathematical works are essentially an encyclopaedia on ancient applied mathematics. Their works in the field of Geometry are still taught to children to this day all around the world.

If $a$, $b$, and $c$ define the sides of a triangle and its semi-perimeter is $p = \frac{a+b+c}{2}$, then its area can be found using the formula $S = \sqrt{p(p-a)(p-b)(p-c)}$. In *Metrica*, Heron proposed rules for the numerical solution of quadratic equations and approximate extraction of square and cubic roots. In the mathematical works of Heron, the rules are most often not deduced, but instead explained by examples. In this way,

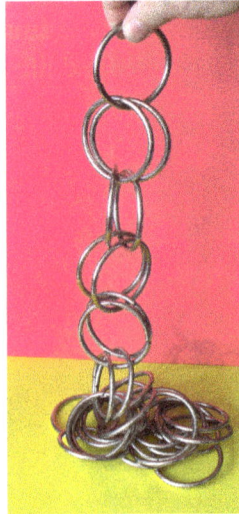

**Figure 1.11**

Heron seems to be part of the tradition of the works of mathematicians of Ancient Egypt and Babylon.

Heron was famous both for their mathematical works and constructions. From their work *Pneumatics*, we can learn about the existence and principles of operation of several ingenious hydromechanical devices of that time. In *Automata*, construction guides and descriptions of temple and theatre automata are given. In Heron's essay *En Dioptra*, the rules of land surveying with the help of a special device called a diopter — the predecessor of the modern theodolite — are set out.

Heron managed to systematise the knowledge of his forebears in the field of optics as fully as was possible. Following their works, scientists began to divide optics into catoptrics, the science of reflection, and dioptrics, the science of changing the direction of light rays when they hit transparent media, or, as we say now, about refraction. In the treatise *Catoptrica* (the study of reflection), Heron justifies the straightness of light rays with an infinitely high propagation speed. Furthermore, they give a proof of the law of reflection, based on the assumption that the path traversed by light should be the smallest of all possible. Following the law of reflection, Heron considers various types of mirrors, and pays special attention to cylindrical mirrors.

By the third century CE, Alexandria had fallen into decline. In the following centuries, its museum, library, temples, and theatres (the centres of pagan philosophy) became the targets of persecution of followers of the new religion Christianity. In 640 CE, the library that had survived until that time was burned down during the Arab invasion of Egypt. The golden age of Greek analysis was over, but some manuscripts have been preserved thanks to the most enlightened caliphs from among the Arab conquerors. In the Middle Ages, following the fall of the Roman Empire, the engineering achievements of Alexandria were preserved and developed in the Islamic world.

In the ninth century CE, the main successors of Heron were three Iraqi brothers from the Banu Musa family in Baghdad. Their book *The Book of Ingenious Devices* contained descriptions of many different mechanisms and their logics, clearly tracing the creative heritage of this amazing genius of antiquity, and shepherding these ideas through to the present.

Currently, thanks to the information revolution and the ease of obtaining a variety of information, interest in the knowledge and inventions of ancient civilisations has received new fervour. An example of this is the physical puzzle that became very common and fashionable among schoolchildren not so long ago, "Heron's chain". The toy is two chains that are intertwined in a special way so that both chains seem to be one chain. If you take such an intertwined chain with one hand for the upper ring, and the other for the second, then, releasing the upper ring, you can observe how it, sliding between the links, will fall to the very bottom of the chain and get stuck on the last link. Then you can again take hold of the upper ring with one hand, the other for the second, and repeat its fall. The cycle can be repeated for any length of time; it is not at all easy to guess that the upper ring does not fall at all, but only the illusion of its falling is created with a simple wave rotation of one chain relative to another.

Like almost all of Heron's inventions, their chain has no practical significance and can be used exclusively for entertainment purposes. But Heron's talent as a mechanic and inventor invariably leads to wonderment.

# 1.11 "Magnetic Olives" (Pertaining to *Energy*)

It would seem reasonable to place this toy in the chapter concerning electricity, as this toy does not only demonstrate the force of magnetic attraction. These neodymium magnets are a fantastic opportunity to discuss a fundamental yet nebulous concept in physics, *energy*!

The use of this toy is not as intuitive as many of the other mechanisms which we have explored. First of all, you need to take two magnetic balls or olives in one hand, and without letting them be attracted to each other, spread them with your fingers at a distance of about a centimetre and hold them so. Then they need to be thrown about a metre in height. During the flight up and the subsequent fall, they will be attracted to each other. Elastically colliding, flying apart at a slightly smaller distance, colliding again, flying apart again. This kind of oscillation will last the duration of the flight until the magnets are again in one's hand. The entire flight will be accompanied by loud sound at the moments of impact, merging into a characteristic buzzing sound. The spectacle is very interesting; it is accompanied by a change in a number of different types of energy.

Nowadays, we know many different types of energy. All types of energy, without exception, are associated with another fundamental

**Figure 1.12**

physical concept — motion. This toy displays six different expressions of energy (two types mechanical and four internal energy):

Mechanical Energy:

• Kinetic energy
• Potential energy

Internal Energy:

• Electromagnetic energy
• Sonic energy
• Elastic energy
• Thermal energy

Perhaps the most obvious of these energies being expressed is that of electromagnetism, which causes these magnets to collide with each other. This energy expresses itself through the unidirectional movement of free electrons in the atoms of ferromagnetic materials, which make up all permanent magnets, including our toy. An artificial analogue of a permanent magnet is a coil of copper wire through which a constant electric current travels. Such a coil is called an inductor or a solenoid, and the electromagnetic energy created by this current is described by the formula $W_E = \frac{L \times I^2}{2}$, where $L$ is the inductance of the coil and $I$ is the strength of the current (measured in amperes).

This toy also expresses mechanical energy clearly, being a body moving in a way which we can perceive through space. Our throwing them is the source of kinetic energy in the example. This energy can be calculated by the formula $W_k = \frac{M \times V^2}{2}$, where $W_k$ is the kinetic energy $M$ is the mass of the magnets, and $V^2$ is their initial velocity. The gravity of the Earth attracts the toy and provides us with an example of potential energy. Potential energy is described by the formula $W_p = M \times g \times h$, where $W_p$ is the potential energy, $g$ is the acceleration of free fall on the Earth, and $h$ is the distance of the magnets from one's hand at the point where gravity turns the vector of momentum towards the Earth. The origin of the gravitational attraction force remains a mystery at present. Which particles cause the action of this force are not yet known, however, the simple logic of all other types of energy suggests that it is in motion of whichever particles that the mystery of gravity is hidden.

Sound, as humans perceive it, is a rapid oscillation between the compression and expansion of matter in the air, with the density of these oscillations determining the pitch. These oscillations can be transmitted through any material; however, the eardrum is designed to be sensitive to these oscillations passing through air. One of the few materials which can enter the ear canal which is not air is water, and it makes the transmission of sound much harder to discern. During their projection, the magnets make quite a loud and distinct sound. The energy of sound waves is calculated in a complicated way. The amount of energy per unit of volume of the air is called the density of sound energy. It is determined by the sum of the kinetic and potential energies of all air molecules in a given volume and is described by the formula $w = \frac{\rho v^2}{2} + \frac{\beta p^2}{2}$, where $\rho$ is the air density, $v$ is the velocity of oscillating air particles, $\beta$ is the compressibility coefficient of the medium, and $p$ is the sound pressure.

The answer to the question of how sound arises is perhaps not entirely obvious, although it is very simple. After an extremely fast collision, the magnets fly apart because the metal from which they are made is very elastic. The sonance arises from the toys returning to their shape after compression from collision (reminiscent of the popping sound of a plastic bottle being blown back into shape after collapse). Elasticity is the property of solid or plastic bodies to restore their shape after removing external pressures; it is caused by the realignment of atoms in crystal lattices (for metals) or by the realignment of molecules in polymer chains (for organic compounds). Springs are a simple example of the mechanism of elasticity, the possible energy from their elastic deformation is described by the formula $E_u = \frac{kx^2}{2}$, where $k$ is the stiffness coefficient and $x$ is the linear size of the deformation. For more complex cases, where there is not a clear axis of compression, volume deformation must be calculated. For a hypothetical rod compressed or stretched along its axis, the energy of its elastic deformation is equal to $E_u = \frac{E\varepsilon^2}{2}V$, where $E$ is Young's modulus (which is to say, the propensity of the rod to resist elastic deformation), $\varepsilon$ is the linear elongation, and $V$ is the volume of the rod. You can read more about elasticity in the description of the toy "Elastic popper".

Thermal energy is not so obviously connected to this toy; however, it is indeed still relevant. During the process of deformation, a small amount of thermal energy is released. The thermal energy of any solid body is the sum of the kinetic energy of the vibrational movements of atoms and the kinetic energy of free electrons. Temperature change over time will be sufficient to calculate this energy. If different substances are given the same number of calories of heat energy, they will heat up to different temperatures as determined by the specific heat capacity of different bodies' constitutive materials. The amount of heat is determined by the formula $Q = cm\Delta T$, where $c$ is the specific heat capacity, $Q$ is the amount of heat, $m$ is the mass of the heated body, and $\Delta T$ is the difference between the initial and final temperatures of the heated body. Another characteristic that affects heating is thermal conductivity (the ability of a substance to conduct a heat flow through itself at different temperatures across its surface). The characteristic of thermal conductivity is the coefficient of thermal conductivity $\lambda$, which varies depending on composition, density, and structure for a particular substance. Denser substances have better thermal conductivity just as dispersed substances are among the best thermal insulators.

# 1.12 "Elastic Popper" (Pertaining to *Elasticity*)

This toy is interesting for its unexpected behaviour. To use the toy, one should push the bulb of the cap in on itself so that the smaller surface side is stretched to be the outermost surface. Spin the toy when dropping it so that it lands on the bulb. It will restore its original shape with a loud click and jump unexpectedly high. The height of this lift will be almost twice as high as the one from which you dropped the toy. This, at first glance, surprising effect is explained by the release of elastic energy. You gave an additional reserve of elastic energy to the rubber body of the toy, turning it inside out.

Several definitions could serve to describe elasticity as a physical concept. Put simply, elasticity is the property of solids to restore their shape after removing the external forces. Given more finesse in its description, elasticity is the property of bodies to restore their shape and volume (for solids) or only volume (for liquids and gases) after removing the external forces. To be precise, elasticity is the ability of matter to resist changes in its shape and volume when exposed to external forces. This ability is due to the result of a change in the balance of two forces that have an electromagnetic origin.

**Figure 1.13**

The first of these forces are intramolecular forces. These are caused by the interaction of electrons on the outer shells of atoms and are responsible for chemical bonds in the molecules of a body. These forces are described in chemistry in detail, but the general principle of the physics of their behaviour is that the outer shell electrons help bind the molecule together and through saturation prevent other molecules forming from the atoms interacting between molecules. A clear example of these forces in action is in $H_2$, the molecule of hydrogen: a third atom cannot pair with the pair of hydrogen atoms as their outer shell electrons have bound to each other exclusively. These forces decrease rapidly over distance. If the electrostatic forces decrease in an inversely quadratic dependence on the distance, then the chemical forces decrease inversely proportional to the seventh power of the distance. Therefore, the main sphere of action of chemical forces is limited by a distance close to the size of the molecules.

The second of these forces are intermolecular forces, also known as Van der Waals forces. These forces bind the molecules of a substance together. Van der Waals forces decrease with distance, are somewhat weaker than chemical forces, and are inversely proportional to the sixth power of the distance between the molecules.

In gases, the distance between the molecules is usually much larger than the size of the molecules themselves, about ten times, but the Van der Waals forces attract them to each other and cause chaotic movement. Due to the repulsion of molecules at their relative convergence, the internal pressure of gases is created. The frequency of collisions determines the temperature of the gas and depends on the speed of movement of the molecules.

In liquids, Van der Waals forces hold molecules at distances close to their own diameter, allowing them to move relative to each other. However, they do not allow the molecules to fly away from the total volume of the liquid, forming surface tension forces.

In solids, the Van der Waals forces are responsible for holding molecules in geometrically correct nodes of the crystal lattice, allowing them to oscillate around a single point only very quickly and within extremely small limits. The distances between the nodes of the crystal lattice in solids, as in liquids, are close to the size of a molecule. As an example, in a cubic crystal lattice of iron, the molecule of which consists of only one atom, the dimensions between the nodes are about 2.866 angstroms (one angstrom is 0.0000000001 m or $10 \times 10^{-10}$ m) and the radius is 1.266 angstroms.

Crystalline solids are not the only bodies to be conditionally called solid; there also are amorphous solids. In these bodies, Van der Waals forces do not press the molecules in a strict order like in crystalline ones. Instead, their structures emerge in diverse ways: sometimes arbitrarily, sometimes in certain chains. In such bodies, the molecules also oscillate continuously and can shift relative to each other. A good example of an amorphous solid is rubber, whose hydrocarbon molecules are spiral chains which can easily move relative to each other. Natural rubber is derived from latex (the juice of the Hevea tree, which was first used by some of the indigenous peoples of South America), this chemical got to Europe after the return of the Columbus incursion. If a sulphur atom is incorporated amongst the rubber molecules, one can get ordinary rubber. This is done through the process of vulcanisation, which is the heating and simultaneous compression of certain chemicals. Another example of amorphous solids is glass, consisting mainly of silicon dioxide or silica [$SiO_2$]. Depending on the presence of other molecules, the properties of glass change very much. If the percentage of silica is almost entirely constitutive, with less than one percent of impurities, then the glass is called quartz; its melting point is very high, 2200°C. If up to thirty percent of the body is composed of calcium, sodium, and/or potassium oxides, typically added in the form of soda [such as $Na_2O$] and/or potash [such as $K_2CO_3$], then its melting temperature lies within 1200°C–1700°C. This is the constitution of an ordinary bottle or window. In any glass at room temperature, the molecules do not stay in place, but move at a very low speed, almost equal to zero. The glass flows incredibly slowly.

Although glass is amorphous, it is impossible to bend it quickly at normal temperature without destroying it. It exhibits the property of elasticity: its shape and volume tend to remain constant. At the same time, crystalline iron can be deformed like plasticine. In engineering, this process is called *volumetric stamping*: an iron billet is placed in a special mould, a punch powered by a piston of the desired shape presses on it with great effort, subsequently the plastic crystal acquires the desired shape.

Thus, amorphous substances, like crystalline ones, can exhibit the property of elasticity, and crystalline ones, under certain conditions, can, on the contrary, become plastic. Moreover, the same substance can exhibit both elastic and plastic properties under diverse conditions. The laws which govern the deformations of bodies under the action of various forces are described by the theory of elasticity, which is based on *Hooke's law*.

In 1660 CE, Robert Hooke found while burdening a thin metal string that its elongation was directly proportional to the tensile force (although only up to certain limits, after which the pattern disappeared, and the deformation became irreversible). This law is described by the formula $F_s = kx$, where $F_s$ is the tensile force, $x$ is the elongation, and $k$ is the proportionality coefficient depending on the stretched material. Nowadays, all steel springs are calculated according to this formula. However, Hooke's law is strictly valid only within certain limits of tension, at which the external forces do not cause stresses exceeding the limit of proportionality (the value of which is different for each material, so-called *limit of proportionality*). If this law is fulfilled with small deviations, at which, after removing the tensile force, the deformation disappears, the load does not exceed *the elastic limit*. In the case that an external force causes residual deformations after removing the pressure, it is said that the *yield strength* has been exceeded. The maximum load that the material can withstand before destruction is *the ultimate strength*.

In addition to destruction under the influence of excessive forces, all bodies can shrink. Compressibility is the property of a substance to change its volume under the influence of a uniformly distributed external pressure. As a body will have properties of elasticity, so it will have properties of compressibility conversely. Gases have the highest compressibility, their volume decreases in direct proportion to the pressure during all-around compression, but only at small values. At large values, compression loses its proportionality and overcoming resistance to compression requires exponentially greater pressure for constant reduction of voluminosity. The volume of our atmosphere at sea level can decrease by more than a thousand times. As an example, at a pressure of 5000 atmospheres, gases from the atmosphere will decrease in volume by 1,163 times. When any gases are compressed, their molecules approach, their speed increases, and significant heating occurs. The processes of compression and expansion of gases obey the laws of thermodynamics, which is described in the descriptions of the toys "Ljubvimjetr" and "Drinking Duck". Solids and liquids are also subject to compression, but to a much lesser extent than gases. Moreover, contrary to a very common opinion, liquids compress much better than solids. For example, water can reduce its volume by 70 times more than steel at the same pressure.

The study that describes the behaviour of materials under the influence of external forces concerns *the resistance of materials*. These studies are of great practical importance: they allow one to theoretically calculate the constitutive components of various structures to determine their strength.

# 1.13 "Resonant Balls" (Pertaining to *Resonance*)

This toy is made from durable, elastic plastic. There is a common axis from which two balls are suspended, each by two spokes of plastic which can twirl freely from their hinging points off of the wand which forms their axis. If you take the toy by the handle of the wand, hold it upright, and shake your wrist rhythmically at a frequency of twice or thrice a second, then the balls will spin around the axis. This is not the most interesting application of this toy, however. If one times oneself carefully, one can engage one of the balls in motion at a time, with the ball on completing a revolution transmitting its motion to the ball before it, creating the illusion that the ball is oscillating between colours.

Resonance is described as when the frequency of the driving force coincides with the frequency of its own oscillations, a sharp increase in the amplitude of the oscillations follows. The appearance of these balls rhythmically oscillating in their rotations is a useful visual representation of this physical phenomenon. Although the resemblance of resonance in the toy does not lead to a noticeable acceleration of the movements of the balls, it is the coincidence of the frequencies of the reciprocating movements of the hand and the rotation of the balls that adds energy to them at

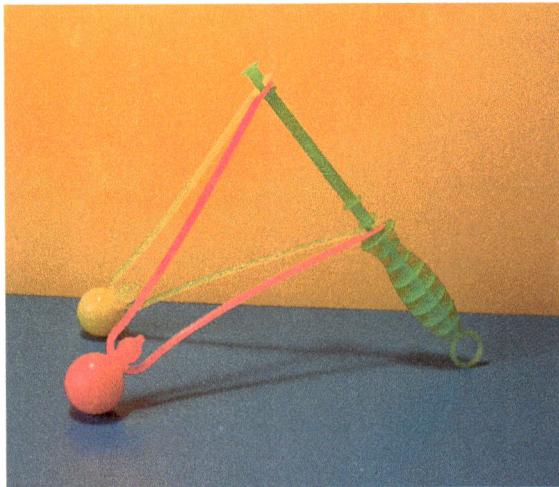

**Figure 1.14**

the right time, compensating for the suppression of acceleration by friction in the axes and air resistance.

Leonardo Da Vinci's hand touches the story of resonance too, with the prerequisites for its discovery attributable to their works from 1492 CE. They noticed that if you hit one of two bells locate next to each other, then the untouched bell beside it may begin to sound too. They also noticed that the same reciprocal vibration property could appear in the strings of a lute.

Galileo Galilei experimentally investigated and described the phenomenon of resonance in their famous work *Discourses and Mathematical Demonstrations Relating to Two New Sciences*. In their experiments, they applied a bow to a tuned violin string such that a thin glass cup began to resonate reciprocally.

The useful applications of resonance are not so numerous as the negative impacts of resonance. It is a property that lends to sonority: it makes it easy to dislocate a car from a parking strip, and it improves the effectiveness of certain power tools. While this may not be an exhaustive list of its virtues, it is a good summation of its practical purposes (one of which was still artistic). Resonance can be a very destructive property as it is about perpetuation and exacerbation. There are several cases of infrastructure devastation from resonance complications, such as bridges swaying in time with the wind until they dismembered themselves. A famous example was the destruction of a suspension bridge across the Tacoma Strait in 1940 CE, which was built to withstand high wind speeds and was left susceptible to resonance with low wind speeds. Similar issues plagued aviation equipment for decades, with resonance having unexpected impacts on aerodynamic mechanisms resulting in "shimmy" in landing gear and "flutter" in the tail. If one has driven a car in wet weather and feels the car start to slide, one must not over-correct, lest they set off an escalating swinging motion in the tail of their vehicle. Resonance is not only an issue in man-made constructions, it also plays into the formation of tsunamis and the mounting of earthquake tremors.

In addition to mechanical resonance, today other manifestations are recognised. In electrical circuits, when the natural frequency of the oscillatory circuit and an external voltage source coincide, an electric resonance occurs: the oscillation amplitude increases sharply in the circuit.

This phenomenon is widely used in electronic radio communication devices, in magnetic resonance stepper motors, as well as for adjusting the power of electrical circuits. The physicist is not alone in seeing resonance in the systems and structures of the world, the term "resonance" has also found a place as metaphor in other spheres of our human life. If a certain event causes a violent response in people, they often talk about a public resonance. When there is a complete mutual understanding between someone, then one may say that there is a cognitive resonance.

## 1.14 "Newton's Cradle" (Pertaining to the Pendulum and the Law of Conservation of Energy)

Of the toys on this list, this one is a particularly iconic feature of science shops and kindergartens the world round. It is made of five balls which are each suspended on two filaments which connect to two parallel rails either side and equidistant from the ball itself. Considering that this instrument was first studied by Galileo, it seems that it would be fairer to call it Galileo's cradle. History is rarely so just.

In 1602 CE, during a service at the Cathedral of Pisa, Galileo Galilei noticed that the chandelier swung from its suspension on the high dome and became fascinated by this motion. Returning home, he started to create experiments with various objects to observe the characteristics of this motion. From this experimentation, Galileo discovered that the pendulum's duration of oscillation was independent from variations in mass at the same length of suspension. To fix time intervals, Galileo then used the frequency of his own heartbeat, which required his creation of the

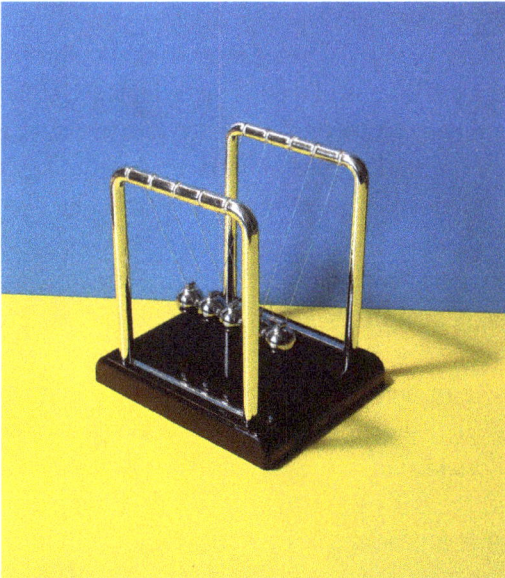

**Figure 1.15**

instrument *pulsilogium*, which entered the medical practice of doctors at that time as a precursor to the stethoscope. The device was a pendulum of variable length which could be set at different suspensions to alter the frequency of the oscillation, and thus could be used to determine the heartbeat of the patient and perceive relative normalcy.

Pendulums were further investigated by the Dutch scientist Christian Huygens. The main goal of these works was to create a mechanical pendulum clock that would be capable of working in rough conditions such as roiling seas. For the Western Europeans living beside the Atlantic, this was a time of significant geographical exploration, so implements of reliable time keeping were crucial. In 1673 CE, Huygens presented their treatise "Pendulum Clocks or Geometric Proofs Relating to the Movement of Pendulums Adapted to Clocks", within which was the description of clocks comprising *a cycloidal pendulum*. They also described the distinction between *a mathematical pendulum* and *a physical pendulum*: a mathematical pendulum is an abstract construction where the suspended body is considered to be a point, the thread is considered to be the thinnest and inextensible, and the suspension is imagined as attached to an ideal point; a physical pendulum is a real solid body that has the ability to oscillate around any of its points that do not coincide with the centre of gravity. Huygens discussed *isochronicity* (the most important property of a pendulum) within their paper as well: they investigated Galileo's observation that the most important property of a pendulum consists in the fact that the time of one of its oscillations does not depend on the mass of the pendulum and the amplitude or scope of the oscillations, and is determined only by the length of the suspension. This turned out to be true only for small amplitudes. They also deduced a mathematical formula for the period of oscillation of the pendulum: $T = 2\pi \sqrt{\frac{l}{g}}$, where $l$ is the length of the suspension, and $g$ is the force of gravity.

This toy made of a chain of pendulums was patented much later, in 1967 CE by the English actor S. Pribble. It clearly demonstrates *the law of conservation of mechanical energy* and *the theory of elastic shock*. By deflecting one of the balls and lifting it to a height $h$ above the table, one gives it a reserve of potential energy $W_p = M \times g \times h$. If the ball is released, then under the influence of gravity, the ball will acquire a certain speed, subsequently the reserve of potential energy will pass into kinetic energy $W_k = \frac{M \times V^2}{2}$.

When the ball that has gained speed descends to touch the first of the chain of stationary balls, an elastic collision will occur. The material of the balls is always chosen to be extremely elastic such that striking against each other can be considered effectively absolutely elastic, thus all the kinetic energy of one ball is transferred to another. This is a manifestation of *the law of conservation of energy*, which states that mechanical energy neither disappears anywhere nor does it appear from nowhere, it only passes from one type to another.

After the collision, the stationary ball, without having time to gain speed, will hit the next ball and transfer its momentum to it. Further in the chain of balls, several of the same collisions will follow, up until the furthermost ball. Since the masses of all the balls are the same, and the elastic metal molecules perform reciprocating vibrations and spend energy only on heating, the furthermost ball after the collision will acquire a speed only slightly less than the speed of the preliminary mobilised ball. Having risen almost to the height of the deflection of the first ball, the last ball will begin to descend similarly to the first and, hitting a stationary chain of balls, will transfer its kinetic energy to them and through them, which will return to the first and then the process will repeat, as will the slight depletion of kinetic energy as heat energy. Such cyclic collisions will continue up to a dozen times until the reserve of potential energy accumulated by the first ball is spent on the work of elastic and friction forces at the suspension points and on the air, after which the balls will stop.

## 1.15 "Pendulum Cart" (Pertaining to Inertia and the Law of the Conservation of Momentum)

This toy is a pendulum on a cart with wheels which has low rolling friction. The way to set this into motion is counterintuitive, but the intermittent character of its movement clearly demonstrates compliance with the law of the conservation of momentum and energy when moving by inertia. Previously potential and kinetic energy were explored in the description of the toy "Magnetic Olives"; law of the conservation of energy was also demonstrated by the toy "Newton's Cradle". In physics, *the momentum of a body* is understood as the vector product of the velocity of a material body by its mass, as explained by the formula $P = MV$.

When considering the law of the conservation of momentum, typically this is best demonstrated by considering a system consisting of at least two bodies. It is assumed that such a mechanical system is insular; typically referred to as a closed system within the natural sciences, meaning it is not exposed to external influences, or within which all non-considered forces are balanced. The law of conservation of momentum in a closed mechanical system states that, in the absence of external

**Figure 1.16**

influences, the sum of the vector products of all the masses of the system of bodies on their velocity remains constant.

This law defines the motion of the pendulum on the cart as follows. One should assume that the toy corresponds to an idealised closed mechanical system that consists of two bodies: a physical pendulum with a swing radius $R$; a mass $M_1$ mounted on a trolley with the same mass, such that $M_2 = M_1$; and that the friction on the bearing of the pendulum is negligible, as it is on the axels of the cart and the wheels upon the table. When the trolley and the pendulum are stationary on the horizontal surface of the table, the centres of mass are located on the same vertical axis.

If, without touching the trolley, the pendulum is deflected by a certain angle a from hanging statically and vertically, as shown in Figure 1.17 (thereby storing a certain amount of potential energy and releasing it), then the trolley and the pendulum will begin to oscillate in antiphase relative to the centre of mass, which will remain stationary. Provided there was no friction, this process could continue indefinitely; the energy in the system will move from kinetic energy to potential energy and back. The cart will go nowhere, thus demonstrating the law of conservation of momentum in a closed mechanical system for this case.

This toy is not limited to this quasi-stationary swinging. The cart can be made to progress haltingly. It will accelerate, then decelerate, stopping for a moment, then moving forwards in the same direction. The way to do this is by setting the cart so that it rests with one pair of wheels against a stationary body. Tilt the pendulum towards the stop at the same angle a so that it turns out to be in the first position as shown in Figure 1.17, and then release the pendulum.

While the pendulum is swinging downwards, the reaction force will press the cart against the obstacle, leaving the cart to remain stationary. As soon as the pendulum passes the second position as shown in Figure 1.17, it will receive an external impulse equivalent to $P_1 = M_1 Vo$ and its inertia

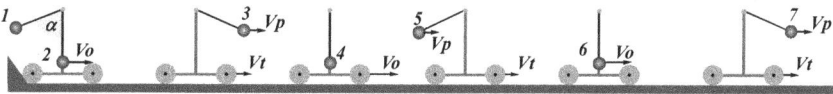

Figure 1.17

will begin to drag the cart along with it, which will accelerate away from the body against which it was set. Simultaneously, the pendulum itself will slow down, and part of the potential energy accumulated when it is raised will pass into the kinetic energy of the movement of the cart and the pendulum itself. Here one should pay special attention to the fact that the closed system received an initial external pulse $P_1$ at the moment when the pendulum was in the second position of F 1.17, coinciding with the centre of mass of the system.

As the pendulum continues to swing, it rises to its third position in Figure 1.17, its speed relative to the cart will become zero. At this point in time, the cart and pendulum will move together on the table with the same speed $V_p = V_t$. Since the masses of the pendulum and the trolley are equal in this hypothetical case, by virtue of the law of conservation of momentum, the absolute value of the speed of their joint movement at relative immobility is equal to half of the maximum speed of the cart, as described by the formula $Vp = Vt = \frac{V_0}{2}$. If the pendulum remained in this position, then the cart would move along with it indefinitely. In this case however, the pendulum will begin to descend, continuing to accelerate the cart until it takes the fourth position as shown in Figure 1.17, transmitting all its momentum to the cart, leaving it stationary relative to the table. At this point in time, the speed of the cart will become equal to the formula $Vt = Vo$; its kinetic energy in its magnitude will be equal to the initial potential energy of the pendulum.

At the next stage of movement of this case, the trolley will be slowed down due to the inertia of the pendulum deviating backwards from the direction of progression. In the ideal case, the pendulum would rise to the original height and will again be stationary relative to the trolley in the fifth position as shown in Figure 1.17. In this position, the speed of the cart will drop to the value described by the formula $Vt = \frac{V_0}{2}$ and the pendulum and cart will again be stationary relative to each other. Continuing from this equilibrium, the pendulum will again begin to descend and apply brakes to the cart until it stops completely now that it passes its sixth position, as shown in Figure 1.17. At this point, one cycle of the movement under consideration will end and the next one will begin the same. Both the pendulum and the cart will stop when the potential energy of the pendulum is transferred completely into the environment, largely by forces such as friction.

If desired, based on the condition of simultaneous fulfilment of the laws of conservation of momentum and energy, it is easy to find the dependences of the speed of the pendulum relative to the cart $V_p$ and the cart relative to the table $V_t$ for any angle of deflection of the pendulum $\varphi$. To simplify the calculations, assume that the pendulum is deflected to a horizontal position at the initial start of the cart, as shown in Figure 1.18.

By virtue of the law of the conservation of energy, the potential energy of the pendulum when it rises to a height $R$ is equal to $MGR$. When the pendulum is in the centre of mass of the system under consideration, all its potential energy will be converted into kinetic energy, as described by the formula $MGR = \frac{MVo^2}{2}$, thus the initial velocity of the pendulum will be equal to $Vo = \sqrt{2GR}$.

At any time, the speed of the pendulum relative to the trolley *Votn* is directed perpendicular to the radius between the pivot and the pendulum. When combined with the speed of the trolley *Vt*, one can determine the vector of momentum of the overall system *Vp*, as shown in Figure 1.18. As the law of conservation of momentum insists, the vector sum of the speed of the pendulum parallel to the plane of the table $V_p(x)$ and the trolley *Vt* are always equal to the initial speed of the pendulum *Vo* within this closed system. Thus, one can write $Vt + VpCOS(\varphi) = Vo$ or $Vt + (Vt + VotnCOS(\varphi)) = Vo$, determining that $Vt = \frac{Vo - VotnCOS(\varphi)}{2}$.

The total kinetic energy of the trolley and the pendulum when the pendulum is deflected by any angle $\varphi$ is equal to its potential energy when the pendulum is first released into free motion, as shown in $\frac{MVp^2}{2} + \frac{MVt^2}{2} + MGR[1 - cos(\varphi)] = MGR$.

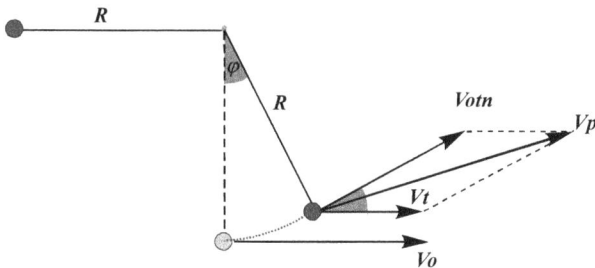

**Figure 1.18**

After several mathematical processes, one can find the value for the speed of the pendulum by the formula $Votn^2 = \frac{4GRCOS(\varphi)-Vo^2}{1+SIN^2(\varphi)}$.

One could easily be mistaken to think that such a method of movement could solve the problem of creating an apparatus that can move in space without interacting with the external environment, due to the forces of inertia during the interaction of its internal elements. At one time, the Russian philosopher and mechanic Ciolkovskij, as well as the pioneer of American rocket technology Robert Hutchings Goddard, were enthusiastically engaged in this problem.

Alas, this problem has no solution. Even if one were to give the load an additional impulse due to internal energy when it interacts with the cart, for example, using a built-in electric motor, one can only increase the speed of the pendulum relative to the cart, i.e., the speed of the cart itself in one cycle of movement. The average speed of the movement of the centre of mass of the cart system will not change at the same time, in the final phase.

Instead, the pendulum and the cart will make oscillatory movements in one place relative to the common centre of mass until the internal energy runs out. Despite the apparent simplicity of the toy, demonstrating with its help the movement of balls by inertia, one can discern the fundamental, unintuitive physical concept of inertia. The term *inertia* in literal translation from Latin, which came to English through French, means immobility. However, today, inertia is interpreted as the ability of material bodies both to remain at rest and to move in a straight trajectory. If these bodies are not affected by force from other bodies, or when acting external forces, with their simultaneous action, balance each other. The metaphysical meaning of the exact wording of this concept often causes disputes and various interpretations, not only of a philosophical orientation, but also in the professional environment of physicists.

Aristotle argued that bodies can move at a constant speed only if a constant force acts on them. After almost two millennia, Galileo Galilei questioned this point of view, having studied the accelerated and uniform motion of bodies, concluding that it is possible to move by inertia without the presence of acting forces. However, talking about the circular motion of our Sun and the planets, they believed that these celestial bodies also move by inertia with no forces acting on those bodies, while today such a movement is explained by the action of gravitational attraction forces. An understanding of inertia close to the modern one was proposed by Isaac Newton in the first of their three famous laws of dynamics, according to which "all matter left to itself has the ability to maintain a state of rest or uniform rectilinear motion". According to Newton's views, the inert properties of any material object are proportional to the amount of matter contained in this object, which is to say its mass.

Although the modern concept of mass has undergone serious changes and differs from the Newtonian one, it nevertheless has not lost its direct connection with the concept of inertia. On the one hand, today mass is determined by the force of gravitational interaction with which a given amount of matter interacts with another mass: this is termed as gravitational mass. On the other hand, the mass of a body can be determined the character of its acceleration under the influence of the interaction force with another mass: this is the termed as inert mass. However, it has been established experimentally, with very high accuracy, that the gravitational and inert masses do not differ from each other in any way and, most likely, represent the same thing.

# 1.16 "Toy Gyroscope" (Pertaining to Gyroscopes)

In 1817 CE, the gyroscope was invented by the German mathematician and astronomer John Bonenberger as a physical device for demonstrating the peculiarities of the rotation of the globe. Their gyroscope was a massive rotating metal ball, fixed in a hinged mount, providing free rotation of the ball around three mutually perpendicular axes. This mount is now known as a "gimbal". It received its modern name from the Italian Renaissance scientist Girolamo Cardano, the inventor of the combination lock; the three-dimensional hinge was only described by them within their scientific works. A similar mechanism was invented by engineer Philo of Byzantium back in the third century BCE in Ancient Greece.

The term *gyroscope* itself originates from Ancient Greek, etymologically derived from *gûros*, translating to a circle, and *skopéō*, translating roughly to observer. In 1852 CE, the term was coined by Abbe Foucault, who used a gyroscopic device to prove the daily rotation of our Earth with the demonstrative aid of their titular pendulum. Foucault measured the angle of rotation of a gyroscope fixed in space relative to a hinge fixed on our Earth turning from East to West with the help of a microscope. The

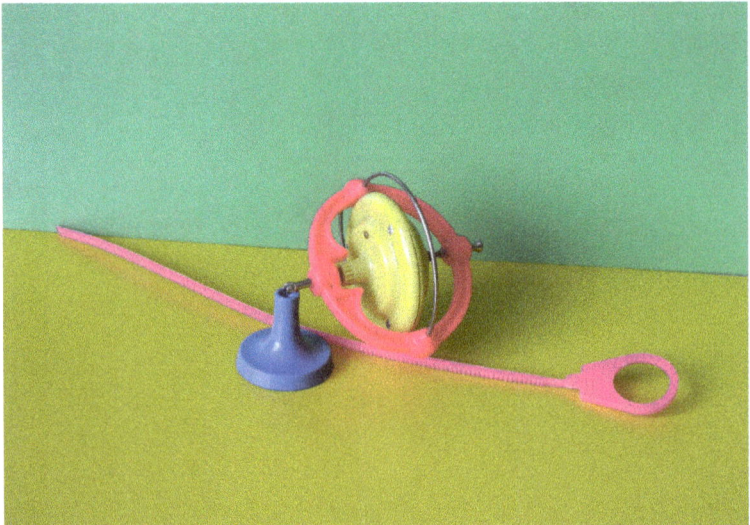

Figure 1.19

use of the microscope was due to the angle of rotation of the gyroscope being extremely small, for an acceptable observation time.

This toy gyroscope is constructed from a small disc-shaped flywheel rigidly fixed on an axis which, with a very low coefficient of friction such as when placed on its pedicle, can stay up right regardless of whether the flywheel's axes of rotation is not parallel to horizontal. To put such a toy into action, one must set the flywheel into rotation by use of a draw cord or some other suited implement. If one were to place it such that the axis of rotation is vertical, the gyroscope will spin like a regular top. If one sets a rotating gyroscope on a stand such that the axis of rotation is horizontal, the gyroscope does not fall to the floor, as one would expect since the centre of mass of the toy is outside the support area. One may be surprised to see the rotating flywheel of the gyroscope will seem to hang in the air and simultaneously rotate around the hinged support on the stand.

The rapid rotation of the flywheel of the toy around its own axis is referred to as *gyroscopic rotation*. The angular velocity of this rotation is determined by how much one spins the flywheel. The movement caused when the flywheel's axis of rotation is not vertical and it is set upon its pedicle, such that the gyroscope rotates around a vertical axis passing through the frame's support point in the stand, is referred to as *gyroscope precession*. The precession rate of the gyroscope $\Omega$ is determined by the formula $\Omega = M/L$ and depends mainly on three physical parameters.

The first parameter to consider is the tipping moment $M$ acting on the flywheel of the gyroscope when it rotates around the fulcrum or suspension equal to $M = mgh$, where $m$ is the mass of the gyroscope, $g$ is the acceleration of gravity, and $h$ is the distance from the centre of mass to the fulcrum.

The second parameter to consider is the moment of inertia $I$ depends directly on the mass of the rotating body, and exponentially more depends on how far its body extends from the axis of rotation. When a small point of mass $m$ rotates at a distance $R$ around a fixed axis, the moment of inertia of this mass is equal to $I = mR^2$. A disc of the same mass and radius, rotating around its centre of mass, has a moment of inertia $I = m^2 R^2/2$. Similarly for a ball of the same mass and radius, the moment of inertia is $I = 2/5$ $mR^2$. In general, to calculate the moment of inertia, it is necessary to divide a rotating body into very small particles, calculate the moment of inertia for each of them, and then sum up these moments. Those familiar with calculus may recognise integration in this process. In the toy depicted in Figure 1.19, since the gyroscope's flywheel is practically a disc, the moment of inertia can be suggested to approach the formula $I = mR^2/2$, and the moment of momentum can be suggested to be $L = m\,R^2\omega\,/\,2$.

The third parameter to consider is the angular momentum $L$, which depends in direct proportion on the angular velocity of the gyroscope's own rotation $\omega$ and on the moment of $L = I\omega$. The angular momentum of the toy gyroscope is the derived from the formula $L = mR^2\,/\,2$. By reducing all the parameters into a single formula, one can determine the angular velocity of precession of the toy to be $\Omega = 2gh\,/\,R^2\omega$.

If one observes the gyroscope in action attentively, one will notice that the free end of the flywheel axis, in addition to describing a circle, also fluctuates slightly, perpendicular to the circle. This suggests that in addition to its own rotation and precession, an unbalanced gyroscope has another characteristic mode of movement. The vertical oscillations of the axis of the precessing gyroscope are called *nutations*.

In addition to mechanical gyroscopes, there are optical gyroscopes based on the wave nature of light. In them, light rays traveling along a rotating circle can form a different interference pattern, which depends on the speed and direction of rotation. Such a picture can be observed, for example, with the help of a rotating coil of fibre-optic cable.

The main field of application of both optical and mechanical gyroscopes today is *navigation*. A wide variety of devices, whose main component is gyroscopes, allow people to determine the location, orientation, and speed of movement of machines quickly and reliably in all imaginable terrains, from the depths of the sea to the far reaches of our Earth's exosphere.

In addition to navigation, powerful mechanical gyroscopes have found application as power units. Thus, for the orientation of satellites in space orbit, it is not necessary to use jet engines with a limited supply of fuel. Having a powered gyroscope of relatively small mass on board, one can use electric motors to spin and rotate the satellite's axis in the desired direction, while the satellite body, due to the reaction force in the support bearings, will turn at the required angle relative to the orbit. In this method of satellite orientation, the onboard fuel supply will remain minimally diminished, and the electricity supply can be easily replenished with the help of solar panels.

Powered gyroscopes are also used on Earth, finding their purpose in the seas and oceans. Control moment gyroscopes help to steady pitching sea vessels. The mass of rotating flywheels in such gyroscopes will be many times greater than in space. They are not necessary for every ship. Basically, cruise liners and expensive yachts are equipped with *gyro-pacifiers*, where the discomfort caused by on-board and keel rolling in stormy weather is extremely undesirable. But their purposes are not only for high-flying technologies and floating casinos; gyroscopes make for a wonderful toy in and of themselves.

# 1.17 "Chinese Top" (Pertaining to the Centrifugal Force of Inertia)

The spinning top is an old toy, found during excavations of sites occupied by people from as far back as the Stone Age. Very similar objects have been found from ancient China and medieval Europe. There are many varieties of spinning tops, most of which have the characteristics of a disc or a cone on an axis. One variation found in China is a truncated sphere with a short handle as an axis, perpendicular to the centre of the circular truncation area. Scottish scientist William Thomson, also known as Baron Kelvin, was the first to become interested in the unusual rotation of the top.

This toy resembles the tumbler investigated previously. It has a stable equilibrium due to its centre of mass being located slightly below the centre of the body. When one spins the top by its handle on a flat surface, it will progressively tilt until the tip of the axis touches the surface. With a jump, it will rotate upright from the tip, and then, already in an upended state, it will continue its rotation on the axis of the handle. Gradually, the

**Figure 1.20**

rotation of the top will slow down due to friction, leaving it to fall over again and (touching the surface with its side) it will cease rotation.

Such a rotation is seemingly counterintuitive. During its initial rotation, the top changes the state of stable equilibrium to an unstable one. The reason for this behaviour of the spinning top is the dynamics of the rotating gyroscope. The behaviour of any spinning top during rotation is determined by several factors. Firstly, it is important how carefully the top is made and how high the degree of symmetry of the distribution of the entire mass around the axis of rotation is. Even the smallest grooves will make the top wobble and will not allow it to rotate for any great duration. Equally important is the shape and material of the top — this influences its inertia and the amount of friction force caused by the air. The greater the inertia of the top, the longer it will rotate. A rougher surface or a larger contacting surface area will slow rotation more rapidly. The ideal amount of influence caused by friction would be a surface area reduced to a fine point of contact. The Chinese spinning top has the characteristics of a rigid sphere and a fine point, but its main feature is that when the axis of rotation is tilted, this point can shift along the surface of the sphere. This occurs at the very beginning of the top's rotation: the centre of mass is shifted from the vertical axis by a small distance and *a centrifugal force* of inertia arises. This force keeps any rotating mass in orbit. This force is directed away from the axis of rotation and is numerically equal to $F = MR\Omega^2$, where $M$ is the rotating mass, $\Omega$ is the angular velocity of rotation, and $R$ is the radius of the rotation orbit.

As inertia builds, the axis of symmetry of the spinning top tilts more and more, such that the centrifugal force of inertia of the spinning top will increase the displacement of the centre of mass from the axis of rotation. The point of contact with the surface on which the top balances, or fulcrum, moves along the surface of the sphere, gradually bringing the plane of the slice to parallel with the surface of contact. When the axis of the handle of the spinning top touches a flat surface, the centrifugal force of inertia $F$ forces the pivot point to move to the axis through the tip; it will rotate on this point just like a classic spinning top. Gradually losing speed because of friction, the spinning top will begin to gradually tilt, eventually resulting in the fulcrum moving back to the sphere from the tip, bringing the spinning top to a halt.

The centrifugal force of inertia is often confused with the *centripetal force*. Both forces simultaneously accompany the rotation of any material body. It is very important to understand that two bodies are necessary for any rotation: one body is needed to force the trajectory of the other to bend. For example, to make a stone spin in a sling, one needs a hand to hold the sling as a pivot as well as for that same hand to impart rotation on the sling with a stirring or swirling motions. A centripetal force directed by the hand acts on the stone, as a centrifugal force acts on the hand itself, which is directed from the centre of rotation. The way that a spinning top resembles the hand spinning a stone in a sling is through the principal forces on display: the role of the proverbial hand is played by its axis through the handle, around which the spinning top's mass, representing the proverbial stone in a sling, symmetrically changes the trajectory of its movement. Centripetal force acts on this body, always directing towards the centre of rotation. Centrifugal force acts on the axis, which is directed in the opposite direction, that is, outwards from the axis of the spinning top.

The Chinese spinning top has another interesting feature. Starting the spinning top to rotate clockwise and watching it from the side, after the fall, one will observe that the top rotates in the same direction: clockwise. However, if one draws an arrow on the top, setting it into motion then stopping it in an upright position immediately after it makes itself upright, one will find that the arrow is inverted and the rotation of the top takes place in the other direction. This is explained by the fact that when the pivot point is shifted, the axis of radial symmetry of the top being horizontal, the top itself does not rotate around this axis when it is set to spinning. Instead, the spinning top always rotates around an axis perpendicular to the horizontal plane, and its fulcrum migrates along the face of the spinning top until it aligns with the axis of symmetry.

# 1.18 "Rubik's Cube" (Pertaining to Rotational Movement)

The history of this toy is interesting. Erno Rubik invented their cube, patenting it in 1977 CE and then setting up production. Rubik was a Hungarian teacher of industrial design and architecture. Subsequently, they became the first official millionaire of the socialist countries of the Eastern Bloc, though the Eastern Bloc would fall a few years after. The inventor's goal was not to create a new toy or entertainment. Rather, they wished, with the help of an unusual puzzle, to increase students' ability in spatial imagination, which is crucial when studying *group theory*.

This very complex mathematical theory is one of the sections of general algebra, and it was developed in the first half of the nineteenth century CE by Evariste Galois. They found a method for describing symmetrical objects mathematically. Such objects are understood not only as purely mathematical abstractions, but also as definitive of material structures from molecules to crystals to intricate shells, even to ideal architectural structures and spiral galaxies. Group theory continues to be intensively studied and developed since its applications have proved to be very productive and in demand in many areas of modern science and technology. Examples can be seen in crystallography, architecture, computer graphics, and many other schools of study.

The mechanism of the toy itself is curious. It has been improved many times and there are many variations of it nowadays. There is a crosspiece

**Figure 1.21**

of three small cylinders intersecting at right angles in three mutually perpendicular planes set in the heart of the toy. To the ends of the crosspiece (with the help of spring-loaded screws) six curved parts are attached, which are like facets of cubes with one of the faces in the form of a concave sphere. The opposing faces of each of the six cube pieces have their own colour. Eight cubes in contact with one of the semi-cubes on the axis are made with protrusions that are not visible from the outside. These eight cubes, being compressed by a cube piece sprung on the axis, form a polygon close to the sphere inside the toy with their protrusions. Since all the parts of the toy are compressed finely by the spring, with a little effort they can move along a spherical surface. By moving the toy parts in this way and changing the colour combinations of individual faces, one can get an unimaginably large number of variations: 43,252,003,274,489,856,000 combinations, which is to say in excess of 43 quintillion combinations. It has been proven that from any such combination, it is possible to bring the toy to its original state, with each face monochromatic, in no more than twenty turns.

It is a curious fact that nature has not endowed any living beings with an organ or a part of the body capable of performing a rotational movement around a fixed axis. Of course, a human hand or an animal's tail can describe circles, but in both cases, there is no rotation in the joint, only a hinge motion occurs. Mechanisms in which there are rotating elements appeared only when they were invented by man and the first of them were drawn carts, millstones, and potter's wheels. Almost simultaneously with their invention, the task of adjusting the speed of rotation, and converting between rotational motion and progressive motion arose.

A decisive role in solving this problem was played by the invention of gears, which appeared in ancient Egypt and transmitted rotation from a water wheel in irrigation systems. The first gears were wooden wheels with cylindrical recesses and pins on the rims. A little later, gears began to be made of metal. Also originating during ancient times, saws with a drive from a rotating water wheel to a crank-driven rod were used when sawing logs, based on their ability to perform reciprocating motion. At the same time, rotational motion was widely used in bow saws and primitive lathes with a handle or pedal drive.

In the Middle Ages, windmills began to turn. Thus, the prototype of the modern propeller was invented; which would be the main aircraft engine until the late twentieth century CE. It is worth recognising again that nobody from amongst the living uses a rotating appendage on an axis for flight, although some plant seeds twirl in flight.

With the improvement and complication of mechanisms, the task of changing the direction of rotation at an arbitrary angle challenges craftsmen and *a gimbal transmission* appears. The rapid development of mechanisation in the factory workshops of the Industrial Revolution era leads to the widespread use of steam engines. Simple and reliable *belt drives* are widely used to transmit rotation and control its speed. Later, they will be replaced by *crankshafts*, *bevel clutches*, and the various types of *gear reducers*: *cylindrical*; *conical*; *worm*; and *planetary*.

Nowadays, rotation mechanisms are far less common. It is relatively rare to find *a flexible shaft* and *a cam mechanism*, *hinges of equal angular speeds*, or *a chain transmission*.

The main, essential, and most common element of any mechanism with rotating parts on the axis are *bearings*: mechanical units capable of providing repeated free rotation along any axis in a fixed support. The main task of any bearing is to reduce the resistance to torque by reducing the friction force, and as a result, reducing the degree of wear of the rubbing parts. There are numerous varieties of bearings which can be disambiguated into five main types:

- friction bearings
- rolling bearings
- magnetic suspension
- hydraulic bearings
- pneumatic bearings

The simplest in design are of course plain bearings, in which the friction force is reduced by lubrication, or a special insert made of a material designed to reduce friction. Their main advantage is their ability to function under large loads due to their surface of contact being minimally diminished. They are also able to take uneven pressure without increased dysfunction, as it must be totally spread to be effective lubrication or surfactant. Their disadvantage is hidden in rapid depletion, as they have so much contact and mobility in their function.

Rolling bearings are divided according to the type of rotation bodies located between the movable and fixed clips. Balls, cylinders, and cones are commonly used to transmit rotation. These bearings are good because, with the appropriate choice of design, they can maintain significant forces directed along the axis of rotation. Another convenience is that their dimensions can be quite large, without requiring a reimagining of their principle of function. The disadvantage of rolling bearings is their high sensitivity to vibrations and shock loads.

The main feature of magnetic bearings is the absence of direct contact between the rotating shaft and the supports — the gap between them is maintained by magnetic repulsive force. There are two different types: active and passive. In active magnetic bearings, the rotor is made of a steel alloy with a high magnetic permeability, and currents are induced in it by electromagnetic coils of complex shape placed in the stator. For the electromagnetic interaction forces to maintain a constant gap between the rotor and the stator, the stator's electromagnets are connected to automatic monitoring and control systems. Such systems are very expensive, and this limits the scope of application of such bearings. Passive magnetic bearings involve the use of permanent magnets in both the rotor and the stator. These bearings have great prospects, but currently only research work is being carried out on these materials. Thus far, results have not been promising.

Hydraulic bearings are essentially plain bearings with an increased volume of liquid lubricant, which is forced by means of compressors and is carried into the gap due to the geometry of the rotating shaft.

In pneumatic bearings, compressed gas plays the role of lubrication instead of liquid. The disadvantage of such bearings is the need for sealing, which complicates the design and maintenance. However, they also have significant advantages over other types of bearings arising from their extremely low friction, absence of destabilisation by vibrations or uneven loads, and a potentially extended service life.

# 1.19 "Astrojax" (Pertaining to the Moment of Momentum)

This toy consists of three identical balls: two of them are firmly tied to the ends of a thin cord, the third ball slides freely along the cord. If you take either of the balls which is at the ends of the cord in your hand and start bringing it up and down, the movements of the other two balls will be unexpectedly complex and interesting. The trajectories of the movement of these two balls flying on a string will be so complex that the problem of their mathematical description cannot be solved by accurate analytical methods. It is possible to predict how the balls will move but only approximately, using numerical calculation methods.

This toy is a deceptively complicated instrument and should not been mistaken by its humble appearance. The acquisition of sharpened skills in controlling the flight of balloons takes not hours or days, but weeks and months. In 1986 CE, Astrojax was invented by Larry Shaw, a graduate student of physics at Cornell University. A year later they patented it, later

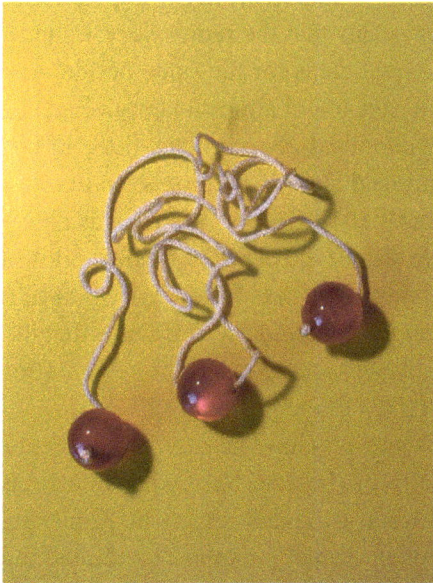

**Figure 1.22**

successfully organising production and sales. The story is very similar to the story of the Rubik's Cube, but Astrojax has only recently begun to gain popularity. Some ambitious youths have managed to master it perfectly and engage in tournaments with each other, much as some youths do with the Rubik's Cube. Astrojax has managed to get into three categories of the *Guinness Book of Records*.

The inspiration for the name of this toy comes from the feature of the two balls rotating around a common centre of mass, much as any two astronomical bodies often do (a famous example being our Earth and our Moon).

The laws of motion which dictate the movement of the celestial bodies in our solar system had been a mystery to humanity for a very long time. Compared to the duration of human civilisations looking up to the heavens, the length of time through which humanity has understood the orbiting, heliocentric model has been very brief. In 1609 CE, an astrologist in Prague known as Johannes Kepler set out two fundamental laws of astronomy in their work *Astronomia Nova*. *Kepler's first law* indicates that the orbits of the celestial bodies follow ellipses, displacing the circles-within-circles models which had been popularised by Aristotelians. Kepler's *second law* states that the radius vector connecting our Sun and any planet sweeps equal areas for the same period. These laws were formulated by Kepler as a result of almost twenty years of work on the analysis of tables of astronomical observations, done in conjunction with the fabulously well-equipped Danish astrologist and proto-astronomer Tycho Brahe. It was not until 1619 CE that Kepler was able to establish a connection between the size of the celestial bodies' orbits and the time taken to complete a revolution. This was followed by *a third law* being stated in Kepler's work *Harmonices Mundi*, which states that the time to achieve a revolution by any planet around our Sun is proportional to the average radius of the orbit raised to the power of three.

Kepler's laws were revolutionary, resolving centuries of confusion about the eccentricity of the celestial bodies and becoming crucial evidence for the heliocentric model, which in Kepler's lifetime was only coming to be accepted. Kepler's method for discerning these laws arose from induction, resting on the tools of kinematics, thus limited by them.

Isaac Newton, developing their laws from the works of Galilei and Kepler, would develop their theory of gravity and use this to define the dynamics between celestial bodies by their laws of dynamics and universal gravitation. In 1678 CE, within *Philosophiae Naturalis Principia Mathematica*, Newton explored through geometric constructions that, under the influence of gravitational attraction, astronomical bodies can move not only in elliptical orbits (circular orbits being a special form of elliptical orbit), but also can follow the trajectories of hyperbola and parabola, much as asteroids and meteorites may do. Later, Daniel Bernoulli and Leonard Euler would work analytically on defining the two-body problem, used to determine the parameters to the motion of two material points attracted by gravity inversely proportional to the square of the distance between those bodies.

Many physicists and mathematicians later paid attention to this complicated problem. In 1795 CE, Pierre-Simon Laplace assumed the possibility that there could be astronomical bodies which are very massive yet do not emit light. Nowadays, a resemblant concept of such an object is a popular topic in astronomy, being *black holes*. Although they have not yet been physically detected with absolute certainty, theoretical research in this direction is actively developing based on the theory of relativity and relativistic concepts. Laplace's conclusion was based on the concepts of classical Newtonian dynamics in the framework of solving a limited two-body problem, when it is necessary to determine the rotation parameters of a mass $m$ incomparably smaller than the mass of the attracting body $M$. Laplace reasoned thus: let the attracting body have radius $R$ and mass $M$; let the small body $m$ rotate in an orbit that almost coincides with the surface of the large body and is held in this orbit only by the force of gravitational attraction. In order for the body $m$ not to fly out of orbit or fall on the attracting body $M$, its velocity must exactly correspond to the formula $V = \sqrt{\frac{2GM}{R}}$, where $G$ is the gravitational constant. This ratio is remarkable because the rate of retention of a small body does not depend on its mass, and therefore this equality can be considered applicable to light. Then, if one assumes the speed of light $C$ to be known, for any attracting body, one can find such a value of its radius $R_g$, at which the light under the influence of gravitational attraction will move along a circular trajectory along the surface of the sphere with a radius $R_g$ and will not have its trajectory diverge from orbit.

In 1916 CE, Karl Schwarzschild fully confirmed the conclusions made based on classical mechanics within the framework of relativistic representations of the general theory of relativity. It was then that they proposed to call *the gravitational radius* the radius of a spherically symmetric material body, determined by the formula $Rg = \frac{2GM}{C^2}$. Physically, this means that the gravitational radius can be determined for any cosmic body and corresponds to the radius of an orbit in which the second cosmic velocity is equal to the speed of light. Assuming the speed of light to be $C = 299.792.458$ m/s, the gravitational constant would be $G = 6.67430 \times 10^{-11}$ m$^3$ kg/s. Knowing the masses of our Earth and our Sun, it would be simple to calculate their gravitational radius.

The mass of our Earth is equivalent to $M = 5.97 \times 10^{24}$ kg, thus its gravitational radius is equal to $RgE = \frac{2\times(6.674\ 30\times10^{-11})\times(5.97\times10^{24})}{(29792458)^2} = 0.00886$ m, or approximately 9 mm.

The mass of our Sun is equivalent to $M = 1.9891 \times 10^{30}$ kg, thus its gravitational radius is equal $RgE = \frac{2\times(6.674\ 30\times10^{-11})\times(1.9891\times10^{30})}{(29792458)^2} \approx 299143$ m, or approximately 3 km.

One can compare the true dimensions of our Earth and our Sun with their gravitational radii:

With a true radius of our Earth of 6,400 km, its gravitational potential is approximately equal to about 9 mm.

With a true radius of about 700,000 km, our Sun's gravitational potential is less than 3 km. If one were to consider the gravitational interaction of not two, but three or more bodies, then there is no strict mathematical solution to this problem. This is called the *n*-body problem, which was defined in 1889 CE by the French mathematician, physicist, and astronomer Henri Poincaré.

# 1.20 "Toy Rocket" (Pertaining to Jet Propulsion)

For the toy displayed above, water is used as *a working medium*. One stores the energy necessary for the flight with the help of a pump and one's own muscles compressing air, which is contained in the small cavity of the rocket above the water. Expanding air pushes water at high speed through the narrow *nozzle* of the rocket. It can be propelled as much as a dozen metres high. There are quite a lot of toys based on the principle of jet propulsion, but the principle itself is unique. This is still the only way to overcome our Earth's gravity.

The principle of jet propulsion is based on Newton's second law of dynamics, which states that the forces acting between bodies are equal and opposite in direction. In a reactive motion, one body repels another, and itself moves in the opposite direction. Nowadays, the most powerful and common are chemical jet engines, which use solid or liquid fuels. Rockets equipped with such engines move by throwing off the combustion products of rocket fuel with speeds reaching up to 4.5 km/h.

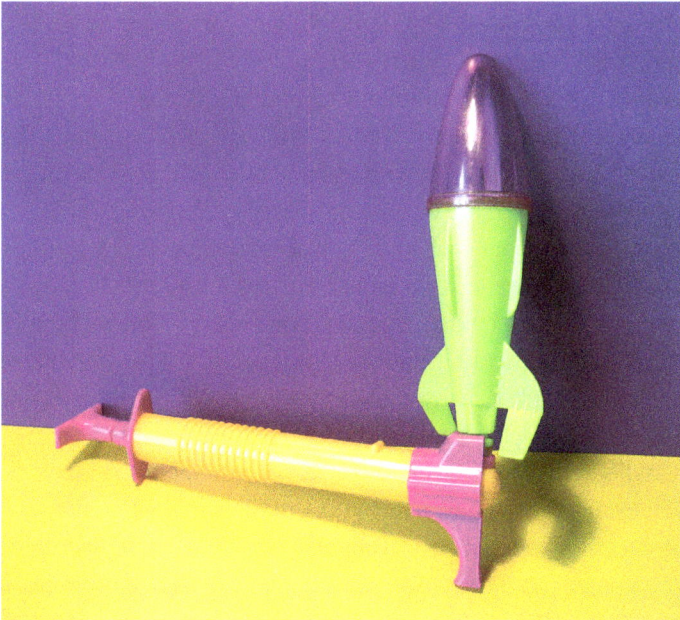

**Figure 1.23**

The reactive principle of movement is not just engineered to the purposes of people. Examples abound within nature: octopodes, squids, jellyfish, and dragonfly larvae all use this. Even some seemingly motionless creatures with shells can move, squirting water to propel themselves, such as scallops.

Nowadays, the theory of jet propulsion is well studied and is described by a study of mechanics *called the mechanics of bodies of variable mass.* In 1897 CE, Mjeshchjerskij Ivan Vsjevolodovich derived the basic formula of jet propulsion. In the further development of this theory (aimed at creating rocket technology capable of space exploration), the main role was played by another Russian, the scientist and inventor Ciolkovskij Konstantin Eduardovich. They invented a multi-stage rocket, which discarded the empty chambers of the body once its fuel supply depleted (what is referred to as stages), having become unnecessary, leaving the functional components to continue their flight with a reduced mass. Without such a method of saving fuel, it would not even be theoretically possible to achieve the minimum speed necessary for a rocket to enter a circular orbit around our Earth. In 1903 CE, Ciolkovskij proposed a scheme for a rocket powered by liquid rocket fuel and theoretically proved its suitability for interplanetary flights.

In 1926 CE, the American scientist Robert Godard and the German scientist Hermann Oberth created and tested such jet engines, putting the theory into practice. In 1933 CE, the first Soviet liquid-fuelled rocket was constructed under the leadership of Koroljov Sjergjej Pavlovich. In 1957 CE, an artificial satellite was put into an elliptical orbit around our Earth with the help of the Fellow Traveller One rocket, created in the USSR on the basis of the R-7 intercontinental ballistic missile. This rocket is known by its Russian name throughout the world: Sputnik. It was the first ever satellite made by humanity and the first of many such achievements of the Soviet Union. The satellite was a metal sphere with a diameter of 58 cm and a weight of nearly 84 kg. A radio transmitter with power sources was mounted inside the satellite, which evenly radiated an intermittent radio signal using four rod antennae in all directions at a publicly available frequency. There was no scientific or other equipment on the satellite, nevertheless, its launch itself heralded the beginning of human exploration of outer space.

The second artificial satellite, called Fellow Traveller Two, was launched by the USSR, just a month after the first. It had a dog named Laika on board with a life support system. This satellite was intended to prove that living beings can live in orbit in zero gravity. Unfortunately, this flight was not built with the capacity for a safe descent to Earth, and although there was a possibility of power supply on board, the dog died in orbit. In 1960 CE, the successful launch of a satellite with living beings on board was carried out by the USSR, when the Fellow Traveller Five spacecraft made seventeen orbits around our Earth with the dogs Bjelka and Strjelka on board. After a flight that lasted about 25 hours, during which several medical and biological experiments were conducted, both dogs returned safely to Earth.

In 1958 CE, the third artificial satellite of our Earth was launched by the Americans; the satellite was called Explorer One. It weighed ten times less than the first artificial satellite, only weighing 8 kg, but had on board, in addition to a radio transmitter and temperature sensors, two more devices. One of them was a Geiger counter: a device for measuring radiation; the other could record the ingress of micrometeorites. During the flight of this satellite, the Geiger counter confirmed the presence of areas with increased radiation, which are now known as radiation belts, in the near space around our Earth. Such regions are formed when our Earth's magnetic field captures charged atomic particles from the depths of space. The hypothesis of the existence of radiation belts located at a distance from our Earth from 5,000–17,000 km was first put forward by the American scientist Van Alen and later they became called "Van Alen belts" in their honour.

In the rapid development of cosmonautics, which began immediately after the launches of the first satellites, the following important stages can be noted:

- In 1961 CE and 1962 CE, the USSR sent the first manned space flights, crewed by Gagarin Jurij Aljeksjejevich on Vostok One and Tjerjeshkova Valjentina on Vostok Six respectively. Pojekhali!
- In 1965 CE, the USSR send the first person to walk in space, Ljeonov Aljeksjej on Voskhod Six.

- In 1969 CE, the first landing on the surface of our Moon by Neil Armstrong and Buzz Aldrin with Michael Collins remaining within the vessel.
- In 1971 CE, the USSR created the first long-term habitable space station. Saljut One was placed into our Earth's orbit, parts of which would become parts of the International Space Station nearly 30 years later.
- In 1998 CE, the International Space Station was created.

The colonisation of our Moon and Mars has been planned, though it is a monumental task compared to any previously undertaken. The implementation of such grandiose tasks will be carried out with the help of already tested and technically proven liquid rocket fuel launch vehicles, but there are projects for other technical solutions.

Even today, there are electric rocket engines which use ions, plasma, electrothermal ejections, and other possible propellants. In all such engines, electric energy is used to accelerate the particles of the working fluid instead of a chemical combustion reaction. The particles of the working medium, which can be ionised molecules of various gases or a substance heated to a plasma state, are accelerated in an electrostatic field or by an electromagnet or in a combination of both fields. Electric rocket engines are simple in design and compact, have a small mass, are easy to operate, and they also do not need complex storage systems for the working fluid. However, along with the obvious advantages of such engines, they have a fundamental drawback: low thrust, which is only tens of Newtons compared to chemical engines capable of developing a thrust of hundreds of thousands of Newtons. Nevertheless, electric rocket engines have their own niche of application. If chemical engines develop huge thrust for a relatively short period of time, then electric engines can work with low thrust, but for a very, very long time. This circumstance puts them out of competition when flying to the most distant bodies of the Solar system. In addition, electric rocket engines can be effectively used by orbiting space stations to correct the parameters of their orbit.

Another promising direction for the development of space technology is the creation of rocket engines that use atomic energy. There are developments of several types of such engines: nuclear, radioisotope, and thermonuclear.

Nuclear rocket engines are jet engines in which the working fluid, most often hydrogen or ammonia, is heated to high energies due to the reaction of a controlled decay reaction of nuclear fuel in a nuclear reactor. Practical work on the creation of such nuclear jet engines was carried out only in the USA and the USSR. It was first started in 1958 CE by the American Orion project. This project was closed in 1965 CE due to the high costs of the Lunar project and the development of the Space Shuttle program. In the USSR, by 1980 CE, the RD0410 nuclear rocket engine was created, which showed a thrust of 35,200 N during tests, but was not used in flying missiles. In 2018 CE, there were reports that the United States is considering the prospects of resuming work on the use of nuclear rocket engines in flights to our Moon.

Radioisotope rocket engines, made with nuclear reactors, do not require the heating of propellant fuels, such as hydrogen or ammonia, to high temperatures for the function, instead using the energy released during the gradual radioactive decay of the radioactive isotopes that comprise the nuclear fuel. The isotopes of uranium U-232 and plutonium Pu-238 are commonly proposed as such substances. Their half-lives are 67 years and 88 years, respectively, during which these isotopes provide a constant and high energy yield. There are models of radioisotope engines in which the products of radioactive decay are themselves used to create a jet. Another type of such engines uses the conversion of the thermal energy from the decay of radioactive isotopes into electrical energy, subsequently using this to power electric rocket engines.

The use of energy of thermonuclear fusion was provided for by the first, theoretically well-developed project "Daedalus", which was developed by employees of the British Interplanetary Society from 1973 CE to 1977 CE. The project involved the creation of an uninhabited interstellar spacecraft with the aim of reaching the star of Bernard, a red dwarf, at a distance of slightly less than six light-years from our Sun. The research project was led by Alan Bond. It turned out that the main difficulty lies in the practical implementation of several technologies necessary for use in the design of the ship: a pulsed thermonuclear rocket engine with inertial plasma retention, a beryllium shield, as well as a supply of helium He-3. Unfortunately, it is still impossible to build such a ship, even if the whole of humanity takes up the pursuit.

Although modern developments of rocket engines using a thermonuclear fusion reaction to heat the working fluid are also far from practical implementation, work in this direction continues. Of particular interest among the latest publications is the thermonuclear engine (Direct Fusion Drive, DFD) being developed in Princeton (USA) by Princeton Satellite Systems. The developers claim that if everything goes according to schedule, the first flight with such an engine can take place as early as 2028 CE.

# Chapter 2

# Liquids and Gases

## 2.1 "Immiscible Liquids" (Pertaining to the Aggregate States of Matter)

This toy is a transparent plastic vessel, intricately and symmetrically separated by transparent walls. The vessel contains three immiscible liquids. One of these liquids is transparent and differs in chemical composition from the other two, which are different colours and denser liquids. The coloured liquids are separated from each other by a transparent wall and when left at rest, the liquids accumulate at the bottom of their wells. If the toy is upturned then, under the influence of gravity, both coloured liquids begin to flow down through small holes in the walls. While falling through the transparent liquid, they do not mix with it, instead travelling through this medium forming teardrops. With the help of this toy, one can extrapolate to consider the properties of the aggregate states of matter, as well as about various combinations of gases and liquids

In this section, the aggregate states of matter that we commonly encounter will be examined. All properties of matter perceived by our organs of perception are derived from the properties of baryonic matter when it aggregates into atoms and their bosons, which form molecules together. There are three aggregate states of matter which we commonly encounter: solid, liquid, and gas. The forces of interaction of atoms and molecules with each other and the speed of their movement relative to each other determine which of these three states the substance is in.

**Figure 2.1**

The solid aggregate state of a substance is characterised by the fact that it always retains its shape, and its atoms and molecules have a relatively small speed of oscillatory motion around central positions that are stationary in space.

In the gaseous state, the given substance is a mixture of individual atoms moving very quickly relative to each other. The forces of electromagnetic interaction and relatively high temperatures will not allow small molecules to attract each other and form larger molecules. The most common element in the universe is hydrogen; scientists have estimated its total mass in our Galaxy as several billion solar masses. Clusters of hydrogen form nebulae out of gas and plasma (which are ionised gases), most of which are concentrated in the region in the centre of our Galaxy. As the temperature decreases, the speed of the atoms and small molecules decreases, and they can form larger molecules. The lowest officially registered natural temperatures of molecular gas is –p271°C or just above 0°K; it was discovered in the young cosmic body known as the Boomerang nebula, which is considered the coldest place in our Universe. About 99%

of the mass of the entire interstellar medium in the Universe and about 98% of the mass of our Galaxy consist of molecular gas, which is a mixture of hydrogen (70% of the total mass) and helium (28% of the total mass), the remaining 2% of the mass of interstellar matter are heavier gases and dust particles of various elements.

Because atoms and molecules move in the volume of any gas at a sufficiently high speed, when two different gases come into contact, they can diffuse into each other. Diffusion allows gases to exist in mixtures with no necessary volume ratio (sometimes these mixtures of gases are termed as gas solutions). The most common example of gas mixtures in people's lives is the air, which is a mixture of nitrogen (78%), oxygen (21%), argon (0.93%), carbon dioxide (0.03%), and several other gases, very small in percentage of volume content, some to vanishingly small amounts. Until recently, it was believed that after the completion of the diffusion process and obtaining a uniform volume mixture of any gases, their separation would not be possible. This turned out not to be the case. Instead, at sufficiently high pressure, some gas mixtures can delaminate. A good example is a mixture of helium and ethylene, which delaminates at a pressure of 100 atmospheres.

It is important not to confuse gas mixtures formed during diffusion and consisting of several gases that do not enter a chemical reaction with complex gases that consist of atoms of several different chemical elements connected by electromagnetic interaction forces. A good example of how this might be ambiguous can be found in the air in three gases. Nitrogen ($N_2$) and oxygen ($O_2$) are abundant in the atmosphere, but nitrogen dioxide ($NO_2$) is a complex gas also found in the air, and despite being constituted by the elements of nitrogen and oxygen, it has different properties. Complex gases can be constituted from many different elements, such as di-fluoro-chloro-methane, which consists of four chemical elements: carbon, hydrogen, fluorine, and chlorine ($CHF_2Cl$).

The liquid aggregate state of a substance shares similarities with both gases and with solids. As with gases, liquids can take the form of the vessel in which they are found. As with solids, liquids have clearly measurable volumes. The relative speed of atoms and molecules in liquids is lower compared to gases, the binding energy of molecules is much greater, and they are more densely situated, which is why there are always surface tension forces at the edge of liquids, creating a barrier that molecules need to overcome during diffusion. These properties of the structure of liquids divide them into those in which the molecules can form solutions when mixed, and those in which the molecules can only mix mechanically, forming emulsions. A good example of a solution of two liquids is a solution of alcohol in water. A good example of an emulsion is when vinegar and olive oil are mixed.

To separate the combined solutions into their constituent components, it is necessary to use special technologies. Some examples of such techniques would be sublimation or the use of membrane technologies.

In membrane technologies, a process called osmosis is used. The foundational idea of osmosis is that as spontaneous diffusion occurs for a solution, if the solution passes through a semi-permeable membrane punctuated by minute pores (with a diameter of 5 micrometers or less) then different constituent components can be separated from each other. An example of such technologies is the extraction of proteins from milk and whey.

During the sublimation process, the solution is heated and, since the boiling point of each component is different, their successive evaporation and condensation occur. An example of such technologies is dividing petrochemical crude oil into its various alkanes and other components.

Unlike solutions, emulsions spontaneously delaminate over time, forming a clear barrier between their constituent components. The time taken to delaminate depends on the chemical composition of the components. Suspensions are another type of solution which are very similar to emulsions; they should be clearly distinguished. Emulsions are a mixture of solutions, whereas suspensions are a mixture of a liquid with insoluble particles throughout. A good example of a suspension is a very fluid cement mortar.

In addition to solutions and emulsions, there are combinations of liquids which do not mix at all. If such liquids are vigorously and forcibly mixed in one vessel, they can form emulsion, but only for a very short duration before almost immediately stratifying under the influence of gravity. This is explained by the fact that the molecules of these immiscible liquids have a sufficiently strong electrostatic potential of the same polarity, and as is known, charges of the same polarity repel each other, whether it is magnets or solutions. Examples of such combinations are:

- Water and glycerine
- Water and fatty oils
- Diethyl ether and glycerine
- Mercury and water
- Fluorinated hydrocarbons (freons) and water.

Two such liquids, most likely glycerine and water, make the toy described above so interesting.

## 2.2 "Erometer" (Pertaining to Thermodynamics)

This toy is very similar to an hourglass, consisting of two glass chambers conjoined by a small aperture acting as a bridge between them. In an hourglass, there is enough sand between the two chambers to fill either to about two thirds of its volume (necessarily not having more contents than either chamber so it is clear when it is finished emptying from one chamber to another). This toy differs in two ways from such a timer. Firstly, the chambers contain a minimally viscous fluid rather than sand. Secondly, the chambers are linked by a hermetically sealed glass tube, which does not quite reach to the bottom of the lower chamber. The part of the tube in the upper chamber emerges at almost half the height of the chamber, where its aperture is shaped like a heart. There are two small holes on the heart, one on the top and one on its lower part, slightly smaller than the hole at the top. This toy's contents are not only this minimally viscous fluid, but also a readily-condensing gas. If the temperature of this gas is the same in both chambers, then the toy is in thermodynamic equilibrium and the liquid level remains unchanged. However, if the lower chamber

Figure 2.2

were to be held in one's hand, then the liquid and gas within would gradually become warmer thus increasing the pressure. Following from the laws of thermodynamics, the liquid would be displaced into the upper chamber.

*Thermodynamics* is a branch of physics that studies the properties of distribution, exchange, and transformations of various types of energy in complex systems and combinations of a wide variety of material bodies. The laws of thermodynamics were discovered in the late eighteenth and early nineteenth centuries by the French experimenters Edmond Marriott, Jacques Charles, Joseph Louis Gay-Lussac, and Guillaume Amontons, alongside their British contemporaries Robert Boyle and John Dalton. The authorship of the discovery of many of these laws still arouses disputes between specialists, thus they are named differently in different countries.

*The first gas law* (Marriott's law, also known as Boyle's law) states that at a constant temperature of a fixed mass, the volume of a gas will be inversely proportional to the pressure under which the gas is (i.e., more pressure, less volume: twice the pressure, half the volume). Mathematically, this law is written as $P_1V_1 = P_2V_2$; the graphic depiction of a volume's dependence on pressure is termed as *an isotherm*. One can easily understand the essence of this law if one imagines a cylinder filled with gas and hermetically sealed with a movable piston. If one slowly moves the piston in one direction or another, the volume of gas in the cylinder under it will decrease or increase in an inversely proportional way to the increase or decrease in the amount of pressure applied by the piston. If the volume of gas increases twice, its pressure on the piston will decrease twice. This hypothetical case might not be reflected perfectly in life: isolating the manifestations of energy transference in a system is never perfectible. One must consider that the piston must be moved slowly to avoid heating the gas, which, with rapid movement, already demand consideration of the second and third gas laws.

*The second gas law* (Charles' law) states that at a constant pressure on a fixed mass, the volume of a gas is determined by its temperature in a directly proportional relationship (i.e., more temperature, more voluminosity: twice the heat, twice the volume). Mathematically, this law is written as $V_1V_2 = T_1T_2$; the graphic depiction of volume's dependence on temperature is termed as an isobar. It is possible to visualise the operation of this law using the same example with a cylinder and a piston, but let the piston move completely freely in the cylinder with the slightest change in the volume of gas under it. If the gas in the cylinder is heated to a certain temperature, it will increase in volume and this increase will be directly proportional to the change in gas temperature.

*The third gas law* (Gay–Lussac's law) states that at a constant volume for a fixed mass, the pressure is determined by temperature in a directly proportional relationship (i.e., more temperature, more pressure: twice the heat, twice the pressure). Mathematically, this law is written as $P_1P_2 = T_1T_2$; the graphic depiction of pressure's dependence on temperature is termed as *an isochore*. For the visual operation of this law, it is necessary to imagine not a cylinder with a movable piston, but simply a sealed vessel filled with gas. If the gas in the vessel is heated to a certain temperature, it will increase in pressure and this increase will be directly proportional to the change in gas temperature.

The above gas laws describe the behaviour of gases for those cases when one of the three parameters in the gas system remains constant. But most often, in real life, all three parameters change simultaneously even in the case of a fixed mass. In 1834 CE, the French scientist Benoit Paul Emile Clapeyron combined all the gas laws to describe such processes, and derived the general gas equation — later this equation also began to bear the name of the scientist. It follows from *the Clapeyron equation* that for a fixed mass of gas, the product of the gas pressure in the system by the volume of gas related to its temperature in the absence of chemical reactions is always constant. This equation can easily explain the characteristic changes of the toy described above.

By covering the lower chamber with one's hand and heating the contents of the chamber, in accordance with the Clapeyron equation, one increases the gas' pressure and consequently its volume. With the temperature of the upper chamber unchanged, the increased volume of vapours in the lower chamber begins to displace the coloured fluid through the tube. This fluid will rise through the tube and pass through the holes in the heart, beginning to overflow into the upper chamber. Since the upper hole is larger than the lower one, the liquid is poured out of it with a small fountain. The warmer the hand, the higher the fountain and the faster the liquid rises. If one waits a little, all the liquid from the lower chamber empties into the upper chamber, the pressure in the upper and lower chambers will equalise and thermodynamic equilibrium will occur. If one grasps the upper chamber with their hand, raising the temperature of the gas within it (consequently cooling the lower chamber), then, in accordance with the Clapeyron equation, the increased pressure in the upper chamber, through the holes in the heart, will empty the liquid back into the lower chamber.

Such heating cycles can be repeated several times, but not indefinitely. The number of cycles will depend on the ambient temperature. When it is cool, there are fewer cycles; when it is warm, there are more cycles. The reader, having carefully read the laws of gas dynamics described above, might wish to consider for themselves why this would be the case.

## 2.3 "Drinking Duck" (Pertaining to Phase Transitions)

This toy attracts and surprises with its seemingly self-motivated motion. It will swing with a flourish without a visible supply of energy from the outside, as if it could do so indefinitely. The toy consists of two hollow spherical glass vessels: the upper one is empty and the lower one is filled with a minimally viscous fluid. The vessels are hermetically connected by a glass tube, with its lower end immersed in the liquid, almost touching the bottom of the lower vessel. In the middle part of the tube, swinging supports comprise a fixed axis. Tilting the toy on this axis, the toy can move such that the upper and lower vessels are on the same horizontal level. The upper vessel is stylised as a bird's head; in the depiction above, it is covered with a fleecy fabric. If all the liquid is only in the lower vessel, the centre of gravity of the glass bird is below the reference points and this ensures the stability of the toy in an upright position. If one were to

Figure 2.3

tilt the toy on its axis, when the upper and lower vessels are on the same level, the stylised beak is lowered into a small container with water on a stand and the fleecy fabric easily absorbs water. In this position, the liquid can flow through a glass tube from the lower vessel to the upper one.

If the toy is initially in static equilibrium, then to make it swing, it is necessary to wet the fabric on the head with water. Quite rapidly, due to the relatively large surface area of the fabric, the water will begin to evaporate. There are six phase transitions to be considered between the three aggregate states of matter:

- Liquids turning into gases, and gases turning into liquids: termed as boiling or evaporation and condensation, respectively;
- Solids turning into liquids, and liquids turning into solids: termed as melting or liquefaction and freezing or crystallisation, respectively;
- Gases turning into solids, and solids turning into gases: termed as deposition and sublimation, respectively.

Any process of phase transition is accompanied by either the absorption or the release of heat. If heat is released during the transition of a substance from one phase to another, such a phase transition is called *exothermic*. The most obvious example of an exothermic phase transition is the combustion of fossil fuels. If heat is absorbed, such processes are called *endothermic*. A good example of an endothermic phase transition is the melting of ice: it is necessary to supply external thermal energy to the substance, heating the solid to weaken the intermolecular bonds, thus allowing the solid to undergo a phase transition into a liquid. Evaporation is also an endothermic process; in order for a part of the liquid to turn into steam, the temperature of the liquid must be sufficient to provide its constituent matter with a speed sufficient to overcome the surface tension forces. Once having broken away from the main body, the matter will pass into a gaseous state. Usually, for matter that is liquid in habitable environments for humans, evaporation is termed as boiling; to remain as a gas, it is necessary to supply energy from the outside of the system all the time. Very moderate temperatures are enough for water to begin to evaporate, molecules break off from the surface and turn into steam. The molecules that fly out of the water as vapour carry away some of the energy thus, in order to compensate for the loss of energy, the body of liquid water absorbs energy from the environment, which is why the evaporation process is endothermic. At the same time, the absorption of external energy by water occurs throughout the entire volume, yet the evaporated molecules carry away energy only from the surface. Consequently, the most superficial matter from the body of water is slightly cooled during evaporation, by about 1°C to 1.5°C.

The same process happens in the toy depicted above. When the water begins to evaporate, the molecules in a gaseous state travel away from the surface of the fabric, the "bird's head" loses some of the heat energy with those vapours, thus the fabric and the glass under it begin to cool and the volume of vapours from the liquid in the upper vessel will decrease. In accordance with the general gas equation (see the Clapeyron equation as described in the description of the toy "Erometer"), a decrease in the temperature and volume of vapours will provide a pressure difference in the upper and lower vessel, due to which the liquid from the lower vessel will begin to rise to the upper one. When a sufficient mass of liquid is in the bird's head, the centre of gravity moves to the position above the supports, resulting in the toy tilting over and assuming a horizontal position. After tipping over, the bird's cloth wrapped beak will sink into the water for a short time and, being saturated, will make up for the loss of the evaporated water. At the same time, some of the vapours of the liquid from the upper vessel will move to the lower one, causing the centre of mass to drop, resulting in the toy assuming a vertical position. From this point, the evaporation process and the rocking of the toy will repeat.

One may get the impression that one can build a perpetual motion machine on the principle of this toy's operation, which will work indefinitely and completely free of charge. However, this is certainly not the case, (more information about perpetual motion engines can be found in the description of the toy "Gravity Cart") in fact, the toy will only work as long as the water necessary for wetting the bird's beak is accessible. At the same time, if one considers water as the "fuel" for the operation of this toy engine, then its efficiency would be so minute that it would not be of any practical benefit. There is a benefit to the toy as a toy nonetheless, it provides the pleasure of allowing one to watch its fluctuations and the opportunity to study the physical principles underlying its operations.

# 2.4 "Cartesian Diver" (Pertaining to Hydrostatics)

This toy is stylised as a creature and is hollow such that it may float. The toy is constructed such that it turns out to be almost completely submerged. The lower part of the internal cavity of the toy has an aperture. The toy can be used to demonstrate its design when it floats within a sealed, compressible vessel. If one were to press their hand on the lid of the vessel or to squeeze the vessel's walls, the pressure inside the vessel would increase. The increased pressure will compress the air above the surface of the water and, acting through the water, also will increase the pressure in the cavity of the toy. Since the weight of the toy does not change, yet its volume decreases, the Archimedean buoyancy becomes less than the weight of the float, resulting in the toy slowly sinking within the vessel much like a diver. When additional pressure is no longer applied to the vessel, the volume of air in the cavity of the vessel and the toy is

**Figure 2.4**

restored, the buoyancy force again overcomes the weight of the toy, and the toy again floats. The only limit to this operation is one's desire to apply pressure.

This toy owes its name to the Latin pseudonym "Cartesius", adopted by the great French philosopher and scientist whose real name was Rene Descartes. It seems unlikely that this toy was only invented by the time of Descartes' notoriety, although the first description of these properties was made in 1648 by Descartes' contemporary, the Italian scientist Raffaello Maggioti, in the scientific work *Renitenza Certissima dell'Acqua alla Compressione*.

In 1643 CE, the Italian physicist Evangelista Torricelli had already proven the existence of atmospheric pressure, and that the air itself has weight. They also conducted an experiment with a glass tube about one metre in length, sealed at one end and filled completely with mercury. Placing such a tube vertically and closing it at the bottom with his finger, he lowered it vertically into another, wider vessel also filled with mercury such that the sealed end was at the top. It turned out that releasing the finger and allowing the mercury to pour freely into the vessel under its own weight did not result in the mercury pouring out completely, rather it would remain in the tube to a height of roughly 76 cm high with an empty space above it. The discovered empty space was then called the "Torricellian void", though now it would be known as a vacuum. The tube filled with mercury is still a close model for some present-day barometers, though they are increasingly unpopular due to the toxicity of mercury.

Torricelli's experiments caused quite a stir, both of interest and controversy. In France, Torricelli's contemporary Blaise Pascal helped arouse said controversy. Pascal was only 24 years old at the time, but they had already marked successes in field of mathematics. They also designed and manufactured the first mechanical arithmometer that could operate with six-digit numbers. Pascal, experimenting with liquids, not only repeated Torricelli's experiments, but creatively modified them, which allowed them to discover a very important hydrostatic effect. A simple experiment with a barrel full of water and a thin but high tube inserted into it helped. While on the second floor of his laboratory, Pascal filled the tube completely, pouring one litre of water into it. To their surprise, the barrel could not withstand the pressure created by such a small amount of liquid and burst. Investigating this effect, they found that if a spherical vessel, in the walls of which small holes were evenly drilled, is filled with water completely and an overpressure is created in it through a thin tube with a pump, then trickles of water will escape in all directions with the same pressure regardless of their relative location, be that from above, from below or from the side. Subsequently, this hydrostatic regularity was called *Pascal's law*, which in mathematical form for a closed volume of liquid is written $\frac{F_1}{S_1} = \frac{F_2}{S_2}$, where $F_1$ and $F_2$ are the pressure force of the liquid falling on surfaces whose areas are equal to $S_1$ and $S_2$ respectively.

Pascal's law is widely used in engineering; all hydraulic presses work on this principle, and this is not the limit of its applications. In a closed hydraulic system, a piston with a small area $S_1$, acting with a relatively small force $F_1$, provides a pressure $P$ equal to $P = \frac{F_1}{S_1}$. According to Pascal's law, the same pressure will be on the other piston of the hydraulic press $P = \frac{F_2}{S_2}$. Since the area $S_2$ is much larger, it follows that for $S_2 > S_1$, thus $F_2 > F_1$, meaning a more significant force can be obtained from the second piston for the same amount of pressure. In addition to the law on the uniformity of pressure distribution in liquids and gases, the formulation of two hydrostatic paradoxes is also named in their honour.

The first paradox states that the pressure force of a liquid on the bottom of a vessel created by its weight may differ from the value of this weight. This indeed is the case. If the vessel has a small area at its base, and its upper part is much wider, as is the case for an amphora, then the pressure on the bottom is very small. However, a vessel standing on a table will not be at all lighter than the liquid poured into it. On walls that have a slope to the surface of the table, the liquid will exert pressure in a direction perpendicular to the walls, but the resulting force of this pressure through the walls themselves will press the table so that the weight of the vessel with the liquid will be the same as if it were simply weighed on a scale.

The second paradox has to do with the pushing force of Archimedes. Traditionally, this force is considered equal to the weight of the volume of liquid displaced by the body immersed in it. However, Pascal noticed in their paradox that a body of a certain mass may well float in a vessel filled with a liquid of much smaller mass. This can be imagined by the example of a large ship that sailed into such a narrow lock that the volume of water between the ship and the walls of the lock is much less than the mass of the ship.

This paradox is explained by the fact that the Archimedean force is not created by the entire volume of the liquid, but only by the column of liquid whose height is above the bottom of the vessel. Even if this liquid is very small, its height will provide pressure that acts on the bottom of the ship and ensures its buoyancy. The main thing here is that the volume of liquid in which the ship floats is closed and when the ship is immersed in it, the liquid rises. This becomes even clearer if you imagine two glasses: the first is narrow and filled with water; the second is only one millimetre wider but filled with water only a centimetre from the bottom. If the narrow glass is inserted into the wide glass, then by its weight the narrow glass will cause water to rise in the gap between the walls of both glasses. If a wide glass is high enough, then a narrow one will be able to swim in it. Thus, Pascal's second paradox to a certain extent clarifies the formulation of Archimedes' law on floating bodies, which can be more correctly formulated as follows: a body immersed in a liquid is affected by a pushing force equal to the weight of the liquid in the submerged volume of the body.

Descartes was intrigued by Pascal's extraordinary way of thinking. Descartes was much older and was already a recognised authority in mathematics, physics, and philosophy. In 1647 CE, on Descartes' initiative, the two contemporaries met in Paris, where they discussed many current problems of their time, including experiments on hydrostatics. Descartes' point of view on the existence of a vacuum was fundamentally different from Pascal's opinion. Descartes believed that the existence of a vacuum is impossible, since the whole world, according to their philosophical views, is filled with minute, weightless particles that are constantly in contact with each other. Furthermore, they had fundamental differences regarding religion.

Perhaps it is with these meetings that the legend (unconfirmed by written sources) is connected, according to which Descartes experimented with a device made by themselves, which was effectively the titular toy of this chapter. The device was a tall cylindrical vessel made of glass almost filled with water, the upper part of which was closed with a stretched film from a bull's bladder, tightly and hermetically tied to the vessel with a strong cord. Descartes used an empty chicken's eggshell as a diver which floated in the vessel. The contents of the egg were previously removed through a small hole in the deeper concavity and a sinker was attached to the same hole, forcing the shell to barely float in the water. Pressing their hand on the film of the bull's bladder, Descartes applied pressure to the vessel, which acted through the water on the air inside the suspended, floating eggshell. Under the influence of additional pressure, the volume of air in the shell decreased, consequently sinking the shell.

Nowadays one can find an interesting modification to this toy, invented by German masters of glass-blowing art in the eighteenth century. Its feature is the ability of the diver to quickly rotate around its own axis. This will happen only if, when diving, one does not constantly squeeze the vessel with their hand, instead pumping the vessel rapidly. The whole trick is hidden in the design of the diver: water under pressure enters its cavity not through one hole located on a vertical axis, but through two asymmetrical holes offset from this axis. When the external pressure is withdrawn, the water coming out of the diver through these two asymmetrical holes creates torque. There are many options for acquiring such a toy; one of them is given in Chapter 5 of this book.

## 2.5 "Submarine" (Pertaining to the Development of Diving Technologies)

This toy submarine cannot move horizontally as it lacks a propeller, but it dives and floats at the will of its operator. This control of the toy is enacted through the fact that the hull of the craft has holes through which water can freely flow in and out. A small balloon with a flexible plastic tube is placed inside the case, the other end of which is hermetically connected to a medical syringe with a volume of 20 ml. Filled with water, slightly loaded with two weights and with the balloon uninflated, the toy sinks to the bottom of its vessel. The toy will stay under water for as long as one may want, but demonstration of this toy's operation requires that one press down on the plunger of the syringe, thus increasing the volume of the balloon, resulting in the boat starting to float. Having stopped pressing on the syringe at a certain point in time, one can make the toy hang motionless in the body of water within its vessel. If one were to push the entire volume of air from the syringe into the balloon, the toy will float on the surface. Then one can pull the plunger of the syringe, causing the toy to resubmerge. One can play this way for as long as one desires. This

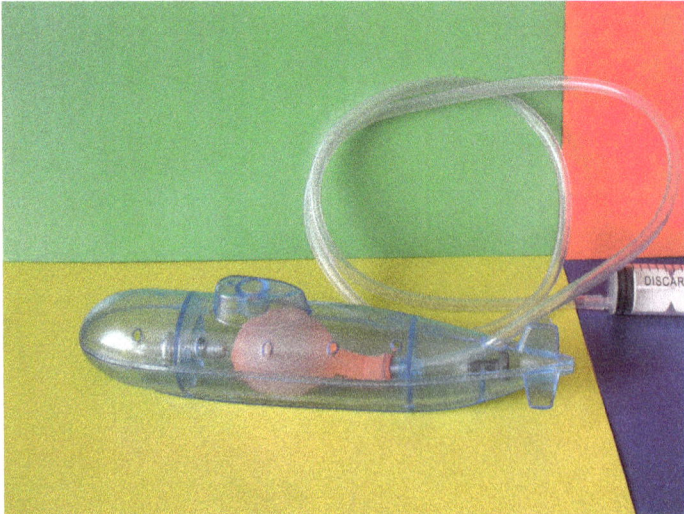

Figure 2.5

interesting toy clearly demonstrates the basic principle of regulating the buoyancy of modern submarines.

The waters of the world are teeming with life, but humanity is not so well suited to living under the water much as we may enjoy living beside it. Most of us like to swim and many of us enjoy diving; the hint of some webbing between our digits, and the presence of webbing in early natal development before the common apoptotic stage, suggests our marine origins. The idea of using something artificially created to facilitate prolonged submersion beyond the bounds of our bodies has long enchanted people. The ancient scientist and historian Pliny the Elder described what was in effect employing a wineskin for this purpose. Hunters for sponges, pearls, and other marine bounties allegedly would tie an additional sac filled with air and dive with it. Staying at a depth of up to 30 m and taking a few breaths from the bag, they could stay underwater several times longer than usual. Another way to increase the time spent underwater, which was considered possible, was to breathe in the depths through a tube lowered from the surface. Practically speaking, neither of these methods are likely possible. The sac obviously would act as a floatation device and would require significant strength to submerge, but this is not the most significant hurdle. As a person travels underwater, the pressure of the water body above them increases by about one atmosphere per every 10 m; at a depth of 30 m, the human body will be compressed by a pressure of about 4 atmospheres. Under such pressure, in accordance with the Marriott law (as has been described in the description of the toy "Erometer"), the volume of a person's lungs is reduced by four times, thus the allegation of the possibility of inhaling air from the surface under such pressure is at best dubious.

The only reliable way to breathe atmospheric air at a depth is to compress it on the surface and transfer the air to the diver through a pressurised hose. In 1819 CE, the English engineer Augustus Siebe first implemented this method using a compressor to supply compressed air to a copper helmet with a glass porthole hermetically embedded in a rubberised diver's suit. With the help of such devices, progressively improved in subsequent years, a person was able not only to stay underwater for a long time, but to perform a wide variety of work. Nevertheless, it turned out that even if one were to work underwater for a short time or simply stay submerged at significant depths for a long time, certain atmospheric constituent gases, mainly nitrogen, significantly increased in the diver's blood. This phenomenon could be extremely dangerous. When surfacing, nitrogen would begin to be released from diffusion in the blood, making the blood fill with minute bubbles that could destroy the walls of blood vessels. Such destruction could be deadly and would be very painful. To avoid such consequences, a method of emerging gradually was invented termed as *decompression*. With this method of returning to lower pressure, the diver would rise in increments of about 2 m to 3 m in multiple stages lasting a few minutes each, until the excess nitrogen would be slowly exuded through the lungs. A diving suit with a breathing apparatus and pressurised air is now considered rudimentary diving equipment, and although it was invented in the nineteenth century, similar equipment is still used today, though there have been significant improvements to the materials themselves and to their safety features. The greatest limiting factor has been the human body, which is not suited to such drastically different climates of pressure. The main instrumental inconvenience of such suits is the need for an air hose, which bonds the diver to the reach of their compressor. This disadvantage can be negotiated with diving equipment whose supply of compressed air is stored in relatively small cylinders attached to the diver's body. There are two types of such equipment.

The first type of such equipment recycles exhaled air, consisting of devices with a closed volume. In this method, carbon dioxide ($CO_2$) is removed from the exhaled air by a controlled chemical reaction, and oxygen ($O_2$) is released during a different controlled chemical reaction, thus being added to the respiratory gases. In 1876 CE, the English engineer Henry Fleuss invented the first device working by this principle. Nowadays, such devices are also used by firefighters when extinguishing certain fires with toxic fumes and in other emergency situations where it is impossible to breathe normal air. These devices can even be run on pure oxygen, which avoids the need for decompression when surfacing, but they become dangerous when diving to a depth of more than 6 m. This danger arises from the fact that when the oxygen pressure in a person's lungs is higher than 1.6 atmospheres, it causes poisoning. This disadvantage can be eliminated by using special breathing mixtures. The main advantages of these recycling devices are firstly, the possibility of a long stay underwater with refuelling periods being extended to up to seven hours; secondly, the absence of exhaled air bubbles, which makes them especially attractive for the military due to its covert potential along with their compactness. The disadvantage of these devices is their complexity, the need for very careful and accurate maintenance, as well as high cost.

The honour of creating the second type of underwater equipment for free swimming underwater, equipment with an open breathing system popularly known as *scuba*, belongs to the French explorer of the world's oceans, Jacques Yves Cousteau. In 1943 CE, Émile Gagnan and Jacques Yves Cousteau invented a special pressure regulator, which from high-pressure air cylinders was able to reliably supply air to the diver at a pressure equal to the pressure of the surrounding water. Aqualungs today are the most common devices for free swimming underwater. The record depth of scuba diving is 330 m. The duration of this dive was only ten minutes and the decompression during ascent lasted nine hours. The maximum duration of a dive with replaceable cylinders, set as a record, is almost six days.

Another direction of the development of water depths is diving in apparatuses, where a person is completely isolated from the surrounding water. Such devices arose from the design of the diving bell, invented in ancient times and which, in fact, is conceptually just a diving helmet upscaled such that the diver fits completely within it. Such devices are still used to this day when it is necessary to deliver a diver in modern equipment to a great depth. The disadvantage of the diving bell is obvious, the need for decompression remains.

The submarine in its modern sense was intended for military purposes, so most of the submarines built earlier and in commission today are combat submarines unfortunately. There are also underwater vessels created for scientific research purposes, as well as serving just for entertainment, but their number is disproportionately smaller.

In 1620 CE, the first submarine was built in England and tested on the Thames. It was intended for entertainment and arranged in a different way. Like modern boats, it was submerged when a special container was filled with seawater, but it surfaced only when it dropped solid ballast, which was an additional solid cargo prepared in advance. Of course, this method of diving was extremely uneconomical but was a product of the possibilities of contemporary technologies.

All combat submarines use the same principle as a toy submarine to adjust buoyancy. In their design, there is always a ballast tank: a hollow container filled with seawater using pumps when diving is necessary. If it is necessary to surface, the water from the ballast tank is ejected from the body of the submarine with the help of compressed air pre-pumped into special cylinders. In this way, submarines can change the depth of its dive while remaining stationary: this is the hydrostatic method of adjusting the depth of immersion.

However, a submarine can also change its position in relation to the water surface with the help of depth rudders, which are similar to airplane wings, only much smaller and able to rotate around its longitudinal axis. Depth rudders work only when the vessel is moving underwater as water must flow around it: this is the hydrodynamic method of adjusting the depth of immersion.

All submarines have a diving depth limit, which is determined by the limited volume of compressed air on board. This is explained by the fact that the greater the pressure is outside the vessel, the greater the amount of compressed gas necessary to displace the ballast water at a constant pressure. Typically, the amount of compressed air pressure cannot exceed between 200 atmospheres and 300 atmospheres, as there is a limit to what pipelines, cranes, and valves can withstand. Sooner or later when diving, the volume of the submarine may become equal to the volume of compressed air, which would make them functionally purposeless. The maximum diving depth of modern submarines rarely exceeds 400 m to 550 m.

Bathyspheres were created at the beginning of the last century to study life at depths not accessible to submarines. By their design, they are a hollow steel ball with strong thick walls and small portholes made of high-strength glass. The spherical shape is chosen in order to withstand huge external loads when the bathysphere is immersed on a strong steel cable into the unexplored depths of the world's oceans. For the breathing of aquanauts, a supply of compressed air in cylinders is provided in the bathysphere. In 1934 CE, the American aquanaut Otis Barton managed to reach the maximum depth of immersion in the bathysphere equal to 1375 m. However, the bathysphere was limited to this depth by the weight of the steel cable, which would have become too heavy were it to descend any further. Such a craft would no longer be able to withstand the load created by its own weight and the weight of the bathysphere, inevitably resulting in destruction.

In 1960 CE, the world record for the deepest dive was set, with the help of another deep-sea apparatus. French scientist Jacques Picard and Lieutenant of the US naval forces Don Walsh designed a special bathyscaphe named "Trieste", which allowed them to reach a depth of 10,919 m. They used this vessel to reach the bottom of the Challenger Abyss, the deepest point of the world's oceans, located in the Mariana Trench in the western Pacific Ocean.

The bathyscaphe "Trieste" was outwardly similar to a small submarine, it had an oblong float about 15 m long, to which a strong spherical gondola (to be occupied by humans) was attached. The fundamental difference between a bathyscaphe and a submarine (being the main characteristic of bathyspheres) is the method of ensuring buoyancy. In all submarines, without exception, the heavy steel hull is kept afloat by the air enclosed inside it. However, at vast depths, the weight of water displaced by the submarine's hull with air will become less than the weight of the submarine. The Archimedean buoyancy force equal to this weight having been overcome, the submarine will begin to sink, no longer being able to surface. In the bathyscaphe, there is no air cavity, and its hollow body is filled with gasoline. As gasoline is incompressible but at the same time lighter than water, it is this substance which provides buoyancy. Through an elastic partition in the hull, gasoline absorbs compression from water pressure during immersion, and the hull of the bathyscaphe turns out to be practically unloaded. Like all submarines, the dive of the "Trieste" began after taking a small amount of water as ballast. For surfacing, a stock of steel and cast-iron shot was loaded into the "Trieste", which was held by an electromagnet during the dive, reminiscent of those first uneconomical ballast systems from the submersibles in the Thames 300 years prior.

One of the promising directions in the development of deep-sea research today is the development and construction of uninhabited and remotely controlled underwater vehicles.

## 2.6 "Clockwork Boat" (Pertaining to Aquatic Propellers)

A wind-up spring motor rotates the propeller, so that the toy boat glides quite quickly on the water, much as a real boat would. In 1681 CE, the English naturalist Robert Hooke proposed using a screw as a propulsion device for ships back for the first time. In 1752 CE, the theory of the propeller was created by Daniel Bernoulli, but they found widespread use only after replacing paddle wheels, which had been the predominant form of water propulsion for steamships for a quarter of a century. Here it is necessary to say that *the engine* and *the mover* are completely different things. A propulsion device is a technical device that, when mounted on a vehicle, can set it in motion due to interaction with the environment. The engine is a source of mechanical energy for the operation of the propulsion system. The most common mover is an ordinary wheel: the engine rotates the wheel and the wheel, interacting with the surface of the road, drives the vehicle. Propellers can also be considered oars in a rowing boat, where the role of the engine is performed by a person — aircraft propellers and ship propellers are also types of propulsion device, reacting to a

Figure 2.6

different medium. In this toy boat, the engine is a spring motor, and the propulsion is driven by an aquatic propeller.

But the first water mover, most likely, was the barge pole, with which a person managed to master water spaces on a free pontoon. With the invention of rafts and boats, the novel propulsion device of the oar was invented, and it became possible to sail away from the reach of the barge pole. The sail was a revolutionary invention in terms of efficiency and possible speed, but the wind did not always blow where it was needed, and often it did not happen at all, or worse, it could be strong to the point of destruction of the craft. Of course, combining the sail with the oars helped in part, but already in the first century CE, the Roman scientist Vitruvius described an attempt to replace the oars with wheels with blades half-submerged in water. It is verifiable that since the fifteenth century in Imperial China, there were paddlewheels employed on numerous ships within military fleets. It is interesting that even then the wheels with blades were placed both on the sides of the ship and on its stern. As a source of energy for the rotation of the paddle wheels, the muscles of a person or domesticated beasts of burden served as the engine.

In 1807 CE, the appearance of reliable steam engines made it possible for the Irish mechanic and engineer Robert Fulton to build the first paddle steamer: *Clermont*. This ship came to a length of 43 m and was put to work making commercial trips. Soon after this pioneering vessel was on the water, the mass construction of such ships began. However, since stormy sea waves often broke the large blades in the wheels, there was no significant development of steamship construction based on paddle wheels. The propeller readily usurped the paddle wheel. However, in the steadier waters of rivers, paddlewheels have proven to be in demand even in the present day. Today, many restored ships with paddlewheels are operated on rivers and lakes for entertainment and in the tourism sector. Moreover, new river ships are also being built, equipped with modern engines and paddlewheels, which have several advantages over ships with a propeller. Firstly, the paddlewheels placed on the sides allow the ship to turn around almost on the spot, and this is sometimes an extremely useful property for tugboats and other transport within ports. Secondly, in shallow or polluted waters, paddlewheels work much better than a propeller.

The first craft to employ the propeller was not a ship, but a submarine. In 1775 CE, schoolteacher David Bushnell built a combat submarine named *The Turtle* in the United States during the American Revolution. *The Turtle* was designed to swim underwater carrying one person for half an hour and was equipped with a bow propeller. This submarine played a role in the war for independence of the United States from Britain, but it was not effective at causing significant damage. In 1776 CE, *The Turtle* was sunk in battle.

The first successful steamship with a propeller was named *Archimedes*; it was capable of overcoming storm waves. It was built in 1838 CE by the English engineer and inventor Francis Smith. In 1837 CE, Smith (who had previously worked as a farmer) filed an application for a propeller that had several turns and built a small boat to test it in practice. During the tests one of the propellers broke down and only one propeller remained functional. To the surprise of all present at the time, the speed of the boat almost doubled, amounting to about 13 km/h. The patent application was corrected, and the steamer already had a metal propeller with several blades rotating in the same plane. By the end of the 1840's, steamboats with propellers almost entirely replaced paddlewheels. Propellers were able to defeat the paddlewheels due to a number of advantages:

- For their rotation, less force is required, thus it is possible to use reliable high-speed, high-torque engines;
- The propeller has all the blades submerged in water, thus the efficiency is higher. Propellers can transmit 80% of its rotational energy into the water, compared to the maximum 55% for paddle wheels;
- The relative weight and dimensions of the propeller for a similar vessel with similar speed outcomes are significantly smaller;
- Structurally, the propeller design is much simpler than the paddle wheel. The operation and maintenance of propellers is much easier than for paddle wheels as well.

Modern propellers are the most common ship propellers. Their dimensions can range from a few centimetres for outboard boat motors, to the size of three-story buildings in the case of the gargantuan propellers equipped by ocean liners. Structurally, the propellers are divided according to the method of fixing the blades. There are propellers with blades fixed to the shaft, the number of which usually ranges from two to four, but there are also multi-blade propellers, the blades of which can number as many as twelve. Such propellers are called fixed pitch propellers. Adjustable pitch propellers are fundamentally different. Their peculiarity is that during the movement of the vessel, there is a possibility of controlled and synchronous changes in the angles of rotation of the blades relative to the axes perpendicular to the axis of the propeller shaft. This feature allows vessels with adjustable pitch propellers to create a large thrust force at low speed, increasing manoeuvrability and allowing one to reduce the size of the propeller while maintaining its efficiency. With these propellers, the vessel can go backwards without changing the direction of rotation of the main shaft. Despite the obvious complexity of production and operation compared to conventional propellers, almost a quarter of all ships in the world are equipped with propellers with rotating blades, and their number continues to grow.

The main problem of all propellers is the little-known physical phenomenon of *cavitation*. At the very beginning of the use of propellers, it was noticed that after reaching a certain speed of their rotation, the vessel cannot increase the speed of its movement. If you increase the speed of rotation of the propellers, they begin to vibrate strongly and experience powerful concussions. Studies have shown that the reason for this is cavitation. By this time, it was already known that the boiling point of water depends on both temperature and pressure, and it turned out that with rapid rotation on the surface of the blades, water creates such pressure that water begins to evaporate at low temperatures. In brief, a cold boiling occurs. The resulting vapour forms bubbles and their subsequent collapse when separated from the metal surface create breaks in the aqueous medium, resulting in an unevenness of its pressure on the blades. This phenomenon, in addition to the vibrational effect, even leads to the destruction of the metal surface of the blades: they become pitted as if they have undergone rapid, severe corrosion. Initially, cavitation was fought by increasing the area of the blades and maximally deepening the propeller in relation to the waterline. The pressure to find a way to reduce the rate of cavitation of propellers saw improvements to the geometry of the cross-section of the propeller blade by the middle of the twentieth century. Supercavitating propellers began to appear, characterised by their sharp incoming edge and a blunt plane. In such propellers, it became possible, avoiding the destructive effects of cavitation, to increase the speed of rotation of the screws by as much as two times compared to conventional propellers. The improvement of the geometry of propellers continues today, and in addition to this method of combating cavitation, not so long ago, another option was added for supplying an air jet to a rotating propeller. Such a jet partially prevents the cavitation process from developing, especially for planing vessels.

Another common ship propulsion system is water jet propulsion, which creates a directional impulse due to the water thrown off by a powerful pump.

In addition to the water propellers described above, there are also some exotic devices.

- Hydrofoils: these structures enter from the bottom of the vessel almost horizontally, at a slight angle and can raise or lower the hull above the water while moving. They resemble the fins of large sharks and whales. During pitching, such propulsion mechanisms serve to give the vessel more stability.
- Cyclorotors: very similar to paddlewheels, but they are placed under the bottom of the vessel and rotate in a horizontal plane. During its rotation, the blades of such a structure also rotate around their own axes parallel to the axis of rotation of the structure itself and, as it were, rake water, creating a driving impulse. Their thrust is not very large, but they will allow the ship to turn on the spot, therefore wing propellers are used mainly in ports, on shunting tugs and similar vessels.
- Hydrojets: expelling water with a jet of compressed air or special combustive fuels.
- Wave propulsion: using the principle of movement on water reminiscent of the motion of a water snake swimming.
- Magnetohydrodynamic propellers: creating a directed flow of seawater, using it as a mobile electrolyte in an electromagnetic coil of special geometry.

Recently, reports have begun to appear regarding the appearance of vessels equipped with paddlewheels with an improved design of movable fastening and a special shape of the blades. It is expected that their use will provide special manoeuvrability and reliability when swimming in shallow water and in icy conditions.

## 2.7 "Sailboat" (Pertaining to Sailing)

A toy sailboat is a reference to the dramatic history of how humans harnessed the wind to travel the world. The first technical means that gave people the opportunity to radically change their way of living, was most likely water transport, which allowed people to cover considerable distances. The first such crafts were almost definitely separate logs tied into rafts, which would have been followed by dugout boats. With the development of tools and technologies, real ships would have emerged from this. The prototypes of sails were an animal skin stretched by the wind, pushing a raft or boat forward. It is believed that about 5,500 years ago the first sail appeared in Egypt.

Initially, the sail was deigned very simply — it was a rectangle made from interwoven papyrus reeds. Later this would be refined into a rectangular cloth made of dense fabric. Employing these straight sails, the ships of antiquity could only sail with the wind, taking the course of *the forewind*. Sailboats employing multiple masts could sail faster with the wind

**Figure 2.7**

blowing at an angle other than parallel with the axis of bilateral symmetry of the boat. This is intuitive; the sails catch the wind, so wind coming in on an angle will not be captured by the backmost sail fully. However, in this case, flat-bottomed ships could be carried away by the wind from their intended course — thus the ships will drift.

In Assyria, and then in Phoenicia, ships were invented equipped with *a keel*. A keel is a horizontal beam which runs down the heart of a boat. This increased the area of the body of the ship submerged in water without increasing its mass. When the wind blew at an angle not parallel to the keel, and with the sail set a certain way, the keel area would have a greater resistance to water pressure than the resistance of the cross-section of the ship. Consequently, the speed of the drifting of the ship would become less than the longitudinal speed. Therefore, even ships with straight sails, provided that they had a keel, could sail at an angle different to the direction from which the wind blew. However, when not following the forewind, ships with rectangular sails could sail at a maximum deviation from the wind of no more than 45°. If it were necessary to make the angle of the ship's direction of movement to the wind quite divergent then the yardarm, being the horizontal beam to which the sail is attached, would begin to rest against the brace which held the mast, thus the sail could no longer be turned to a more obtuse angle.

This problem was solved by the invention of oblique sails with a shape closer to a triangle by the Arabians. Independently of them, the Indians similarly created such an invention in the eighth or ninth centuries CE. Such sails are easier to raise and lower, requiring far less equipment for control. Most importantly, the rigging of the oblique sail allows one to install it at any angle to the centreline of the vessel. When the oblique sail is installed at an angle ranging from 5°–10° to the wind, it begins to work as an aerodynamic wing and the ship can sail towards the wind. The best sailing ships with oblique sails develop their maximum speed at an angular course, running against the wind by a deviation of 30°–35°.

It is not necessary to sail in one direction when the wind blows constantly in said direction — it is even possible to sail into the wind. To maintain the necessary direction, one can perform tacking. *Tacking* is where one alternates the position of the sails so that the wind blows first to one side, and then, after overcoming a certain distance, to the other.

Thus, on average, the direction of movement of the ship will be in line with what is needed by the crew. The distance that a sailing ship passes from one turn to another, and the timing of these turns, is not at all easy to guess correctly. This is the skill of captains and helmsmen on sailing ships. However, on modern sailing yachts, this task is often relegated to automated course calculation systems.

## 2.8 "Glider with a Rubber Band for Propulsion" (Pertaining to Aerodynamics)

Gliders are heavier-than-air aircrafts that do not have propulsion systems. There is a well-known attempt to design and build a flying model of a bird in about 400 BCE by the ancient Greek scientist Archytas of Tarentum, a contemporary and friend of Plato. How exactly it was launched into flight and its design has been lost to the sands of time. Some sources say that Archytas used a jet of compressed air, others mention flapping wings, but everyone agrees on one thing: their light pigeon, made of wooden parts, could fly at least 20 m. Another interesting wooden model of a bird, whose age was about the same, was discovered in 1898 CE during excavations in Northern Egypt. This model, with a wingspan of about 15 cm and a tail in the form of a vertical keel, is very similar to modern aircraft and is quite capable of briefly planing.

In the old Russian chronicles of the sixteenth century CE, one can find a mention of the slave Nikita who constructed a set of wooden

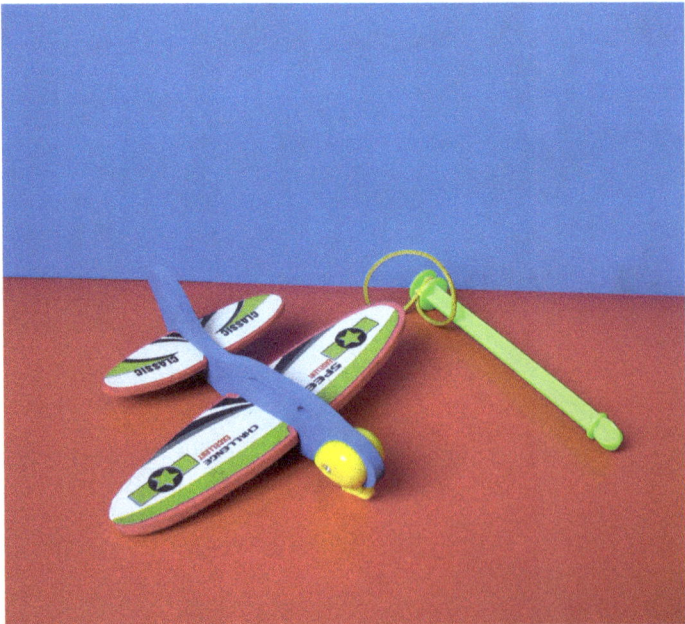

Figure 2.8

wings — sometimes they have been affectionately referred to as Ivan the Terrible's Icarus. They jumped from the Crucifixion Bell Tower in Moscow, with the help of these wings they managed to fly quite a long distance, making it over the fence of the cathedral and landing safely.

In 1891 CE, the first European experience of creating a glider capable of reliably ensuring human flight was carried out by the German engineer Otto Lilienthal, who made more than 2,000 successful flights. Later, they died while testing a new device which would be termed as a biplane nowadays. It was a glider with two wings placed one above the other. Successful gliders were built in 1912 CE by the Russian aviator Arceulov Konstantin Konstantinovich. A rapid surge in the design and construction of gliders occurred after World War I in Germany, where, according to the Versailles Peace Treaty, the construction of aircraft with engines was strictly limited. Nowadays, gliding is a very popular sport and recreational activity. World gliding championships are held once every two years.

Gliders can be launched into flight both from the ground and with the help of *a towing aircraft*. Gliders can be launched from the ground in three ways:

- Using a powerful and fast winch with a cable which is typically diesel powered;
- By towing behind a ground vehicle such as a car;
- Launching with elastic cords.

In flight, the frame's wings are affected by the aerodynamic force created by the incoming air flow. There are two forces to be mediated: drag and lift. The lifting force arises due to the special shape of the wing profile, which is a strongly elongated drop with a slight upward deflection. This shape causes the incoming air to flow around the upper and lower surface of the wing in different ways. Above the wing, it must cover a greater distance than under the wing in the same amount of time. Consequently, the speed of the incoming air above the wing is greater than under the wing. This circumstance is why during flight, there are fewer air molecules above the wing than under the wing, such that the air pressure above the wing is less than under it. In 1838 CE, the Swiss physicist Daniel Bernoulli discovered the dependence of steady gliding on the air flow being laminar (continuously moving without vortices disturbing the medium) to achieve lift. Subsequently, the inverse dependence of the pressure value on the flow velocity of a gas or liquid in a stationary flow was termed as Bernoulli's law. Mathematically, Bernoulli's law for a liquid is described by the equation $\frac{\rho v^2}{2} + \rho gh + p = const$, where $\rho$ is the density of the liquid, $v$ is the flow velocity, $h$ is the height of the liquid, $g$ is the acceleration of gravity, and $p$ is the pressure of the liquid. The first component of the equation, $\left(\frac{\rho v^2}{2}\right)$, is the kinetic energy contained in a particle of the fluid flow, it is called *the velocity head*. The second component of the equation, $(\rho gh)$, is the potential energy of the same particle, created by the weight of a column of liquid or gas with a height $h$. The third term, $(p)$, is the statistical pressure exerted by the liquid on the barriers limiting the flow. For a horizontal air flow, its potential energy is assumed to be zero, and the Bernoulli equation is as follows, $\frac{\rho v^2}{2} + p = const$.

The amount of lift is significantly influenced by the angle between the incoming air flow and the longitudinal axis of the wing profile, though the shape of the wing profile plays an important role as well. The optimal profile and "angles of attack" of the frame can be calculated theoretically, but this calculation is quite complex and is based on the aerodynamic coefficients obtained experimentally. Therefore, any project of a glider or an airplane must pass the stage of testing as a scale model by launching it in a purposefully designed wind tunnel before being realised in full size.

Gliders are quite similar to planes with rigid wings. In 1948 CE, Francis Rogallo filed an application for the invention of a glider made of canvas stretched on a rigid frame. The frame consisted of three thin aluminium pipes, and the canvas had a triangular shape. Nowadays, such a glider with a flexible wing is known as a hang glider. The advantage of such a glider is its low weight and ease of manufacture. Recently, inflatable balloons have been used as a solid frame for a flexible glider, as for example, in the increasingly popular sport *of kite surfing*.

## 2.9 "Toy Airplane with a Push Propeller" (Pertaining to Aircraft)

The toy plane in Figure 2.9 is an amazing example of trying to combine the incompatible. The appearance of the toy quite accurately copies a modern passenger plane with four turbine engines under the wings and one push propeller in the tail section (where it deviates most markedly from the object which it represents). The combination of two fundamentally different types of propellers is absurd from the perspective of practical implementation in real aircraft. However, such a toy is good impetus to investigate the aircraft engines that allowed people to move quickly, reliably, and widely through the sky.

An engine is a technical device that converts a particular type of energy into the movement of a working mechanism. In modern engines, chemical or electrical energy usually is the input. Chemical energy is used in combustion engines and jet engines; electrical energy is used in electric motors. The movement of the working body is most often the reciprocating movement of pistons in internal combustion engines or the rotation of

**Figure 2.9**

a shaft in electric motors. However, there are other types of engines. Pneumatic engines operate at the expense of compressed gas as their input; hydraulic engines use the kinetic energy of flowing water as their input, both are very common in various industries. Piezoelectric devices work by expanding special materials when an electric current is applied to them. Currently, two types of engines are used in aviation: internal combustion and those operating on the principle of jets.

The first aircraft capable of performing controlled flight employed an internal combustion engine. In 1903 CE, the brothers Wilbur Wright and Orville Wright designed and built this first aircraft. Their plane, called "Flyer One", could fly 40 m at an altitude of several meters in 12 s. In 1884 CE, the Russian naval officer Mozhajskij Aljeksandr Fjodorovich designed and built an aircraft with a steam engine and three pulling propellers. Unfortunately, during tests, the power of the steam engine was not sufficient, and the Mozhajskij plane could not take off. The Wright brothers' plane was already equipped with an internal combustion engine fuelled by gasoline. Another fundamental difference was their use of two push propellers. In the future, the Wright brothers carried out significant improvements to their aircraft related to improving flight control and increasing the flight duration and altitude. Between 1905 CE and 1915 CE, similar light aircraft of various aerodynamic schemes were created.

*Monoplanes* are aircraft with two wings located in the same plane on both sides of the fuselage of the aircraft. Biplanes are aircraft with two bearing planes located on both sides of the fuselage. The planes are box-shaped schemes with several rows of bearing planes, resembling ancient Chinese kites. In all schemes, push propellers and pulling propellers were used, with push propellers being preferred until 1916 CE.

This preference was mainly because at the first stage of the development of aviation, its applications were predominantly military. The combat use of aircraft consisted of aerial observation of the combat situation on the ground, bombing and air battles, which are commonly termed as dog fights. The pilots of the first planes fought with the help of small arms mounted on the fuselage, often pistols or machine guns. They did not know how to shoot through a rotating propeller then, which is why the propeller was usually located behind the pilot. In addition, such a propeller would not create vortex pressure on the surfaces of the wings, which greatly facilitated handling. In 1914 CE, a firing synchroniser was invented: a mechanism that combined the impact of the firing pin of a machine gun with the gap between the blades of the propeller, which made it possible to shoot forward through an active pulling propeller. The advantages of the pulling propeller, as it is commonly believed, exceed its disadvantages not only when used on military aircraft, but also on civilian ones. Even though the pulling propeller creates air flow vortices that negatively affect the leading edge of the wings, the point of the pulling force created by the screw is located ahead of the centre of mass of the aircraft, which greatly simplifies the control of the aircraft in flight.

The development of aviation with reciprocating internal combustion engines and pulling propellers reached its highest point by the end of World War II. The limitation is that if the propeller rotates so fast that its blades move at a circumferential speed close to the speed of sound in the air, which is 1,224 km/h (being the speed of sound times pi), air resistance increases sharply. Air particles can no longer be thrown back at speeds that allow the aircraft to develop a speed of more than 650–700 km/h, depending on the geometry of the fuselage. The task of overcoming this speed limit was solved by installing jet engines on aircraft.

The principle of operation of a jet engine is to discard the incoming flow of ambient air accelerated during the chemical reaction of fuel combustion in the closed volume of the combustion chamber. Although there are numerous varieties of such engines, two forms predominate, ramjet engines and turbojet engines. In ramjet engines, as the name implies, the air increases the pressure directly in the combustion chamber and is thrown back at high speed through a nozzle of a special shape that plays the role of a mover. Structurally, these are the simplest engines from the jet family, but they have not found application in aviation. Ramjets are too taxing on fuel, so their scope of application is disposable combat missiles and unmanned drones. In 2018 CE, it was reported that Russia has created a combat missile with a nuclear power source that ensures the operation of a ramjet engine.

Turbojet engines are the most widely used in aviation. The principle of operation of a turbojet engine is that the surrounding air, before entering the combustion chamber of the engine, rotates the turbine of the compressor, compressing the same air. In the combustion chamber, the already compressed air is enriched with atomised fuel, and a jet of combusted products is ejected under high pressure through the jet nozzle. The classic propeller has also found its use in turbojet engines. It is used in turboprop jet engines. In these engines, the jet of combusted products is not just ejected through the nozzle, but also rotates the propeller, thus creating the main driving force; the jet thrust is additional here. Turboprop engines are much more economical than turbojet engines, but they can ensure the flight of aircraft with significantly lower speed. In 1955 CE, the fastest turboprop airliner, the Tu-95, was created by the USSR; its four turboprop propulsion systems allowed it to reach a speed of 925 km/h.

Although turboprop-powered aircraft are now considered obsolete, their modern modifications are increasingly beginning to be used for regional transportation at distances of up to 800 km. But the main array of air transportation is dominated by using turbojet engines, particularly their latest manifestation: the turbofan. These are dual circuit jet engines, in which the air passes through two wide bore pipes, one of which is inserted into the other. The gas turbine of the inner pipe rotates not only the compressor but also the fan, which drives air through the outer pipe, significantly increasing efficiency.

One of the most promising areas for further development in aviation is likely to be the creation of aircraft with electric engines, as well as the construction of new supersonic airliners. These would be similar to the Soviet Tu-144 or the European Concords that flew in the last century, but with improved flight characteristics.

## 2.10 "Ornithopter" (Pertaining to Avian Flight Techniques)

This radio-controlled toy with a lithium-ion battery is a relatively recently invention. A small, but quite powerful electric motor made with neodymium magnets causes light wings to vibrate and the small foam bird begins to fly. The altitude of the flight is determined by the frequency of the wing flaps, and the direction is determined by the turning of the vertical rudder with the help of a tiny electromagnetic coil and a magnetic core. The toy can fly quite high and far compared to its size, responding to radio signals from the remote control from as far away as fifty metres.

The flight of winged creatures is far from a rare sight for most people. Birds and flying insects surround us almost everywhere in the world where people can be found, often throughout the year. Watching their flight, for millennia humanity dreamt of making similar wings for ourselves and using them to rise into the sky. In ancient Greece, a legend was born about the Greek engineer Daedalus and their son Icarus, which tells

Figure 2.10

about an unsuccessful attempt to fly away from the island of Crete on a wooden bird with wings made of wax and bird feathers. Among the drawings and sketches of Leonardo Da Vinci dating back to the years between 1485 CE and 1487 CE, one can find a diagram of an apparatus with flapping wings, which, using the muscular strength of a person, was supposed to allow them to fly as a bird would.

In the seventeenth century, Robert Hooke concluded that the muscular strength and energy given by nature to humans was insufficient for successful flight in the style of birds. Later, this assumption was confirmed by many scientists, but attempts to create an aircraft with flapping wings (what shall be referred to as an ornithopter) were continued when the idea of using additional energy sources appeared. Experiments were conducted on the use of steam energy and other combustion engines, attempts were made which tried to adapt wings to balloons, and so on. However, successful attempts to learn how to fly did not transpire until the beginning of the twentieth century. In 1903 CE, the Wright brothers lifted an airplane with fixed wings into the air. It was in this direction that the rapid development of aviation followed in the future. Nevertheless, the idea of avian-inspired flight was not forgotten. Nowadays, ornithopters have begun to confidently master the airspace thanks to the new capabilities of modern compact, energy-intensive engines and high-strength composite materials. So far, this is happening only in drones. The superiority in energy efficiency of those living beings endowed with the capacity to fly, with their manoeuvrability and variety of flight styles, is still indisputable compared to the entire variety of aircraft created by humanity. This is probably why the further improvement of aircraft based on avian inspired flight is recognised today as one of the most promising areas of aviation development.

The main problem in creating new designs for ornithopters with increased operational properties is the lack of reliable models of avian inspired flight. In 1898 CE, Zhukovskij Nikolaj Jegorovich developed the theoretical basis of aerodynamics for fixed wings affixed to one's back. This foundation remains relevant to these studies to this day. The turbulence that is inevitable during avian inspired flight (unpredictable vortices of the medium through which one flies), due to the effectively infinite variety of vortex flows, still does not allow us to find a single approach to the numerical determination of the parameters of the air environment around the wings. For example, modern calculation methods state that the May beetle, with its mass and aerodynamic imperfection, cannot fly. And yet this creature, as with many even more ungainly insects, flies perfectly well, being able to hover in the air and cause a feeling of envy among aircraft designers. Nevertheless, improvements both in methods of statistical mathematics and in numerical calculations on a computer still leaves us with hope that one day people may be in the sky amongst the birds, flying as they do.

## 2.11 "Rainbow Pinwheel" (Pertaining to Our Earth's Atmosphere)

This aerodynamic toy attracts attention with the continuous rotation of its multi-coloured blades even when the breeze is only faint. Even in the complete absence of wind, one can make the toy rotate quickly by running with it in your hands or spinning on the spot, creating eddies of wind for the blades to capture.

A toy propeller is designed such that it would easily be integrated into the design of an anemometer, being a device that serves to measure wind speed. Accurate determination of wind speed is important in many practical contexts, for instance at airfields for the purposes of landing and launching aircraft. In flight, knowledge of weather conditions is integral for choosing the most favourable route in order to economise on fuel. For meteorologists, the speed of air flows is one of the determining factors in making weather forecasts. Nowadays, on the seas and oceans, there are few sailing ships for which changes in the wind situation are a matter of life and death, but for modern ocean liners, accurate measurements of wind speed help to anticipate safer and more economical routes. Even in

Figure 2.11

closed rooms, knowledge of air flow velocity is always considered in the design of air duct and ventilation systems. An example of how important this can be is in poultry farms, where a failure to meet the required air flow rate blowing upon eggs in incubators can destroy all future chickens in two hours, thus the loss of chicken lives number in the tens of thousands even in a single batch.

Industrial anemometers are very diverse in the construction which underlies their operation. Mechanical anemometers are arranged according to the principle of the toy depicted above; some are even superficially similar. The most common devices where air flow causes rotation are not built with flat blades, but rather light thin-walled hemispheres facing their concave side to the wind. The hemispheres are fixed on an axis, the rotation speed of which is fixed by electronic automation.

There are thermometric anemometers in which the wind speed is altered depending on the degree of cooling by the air flow of a special spinning fan from its constant heating.

Ultrasonic anemometers are characterised by a high degree of sensitivity, using the feature of high-frequency sound that can change its pitch, thus used to determine alignment or deliberate misalignment with the wind's direction. These are more sophisticated devices, used more often in automatic systems.

To measure the speed of air currents in the high-altitude regions of the atmosphere at elevations from sea level between 35 km and 40 km, radio probes are used. These are impermeable balloons (filled with hydrogen or helium) attached to a complex of automatic measuring instruments and radio-transmitting equipment. However, this method has a significant drawback: it is too expensive to launch probes in large amounts or very frequently, which is worsened by the fact that they cannot be used for great amounts of time before their balloons deplete. Consequently, it is impossible to track how the atmosphere changes over extended periods of time with the help of radio probing. Remote measurement methods help to solve this problem.

Wind speed can be measured at a distance with the employment of optical anemometers. In 1844 CE, the Austrian Christian Doppler discovered the physical phenomenon on which these devices base their operational principles. Their discovery was subsequently named after them, becoming popularly known as the *Doppler effect*. The essence of the effect is that when the wave radiation source moves in any medium, the frequency of the emitted waves and the frequency of the waves perceived by the stationary receivers does not remain constant, but changes depending on whether the wave source is approaching or moving away. When approaching, the frequency increases, and when diverging, the frequency decreases. When the distance between the wave source and the receiver is zero, the frequencies of the emitted and received waves are equal. A popular example of this effect is when a person standing still hears the sound of a car passing by with a continuously sounding horn: at first, the pitch of the sound becomes higher and higher, and then, when the car passes the listener, the pitch of the sound becomes lower and lower. In brief, the Doppler effect has found application in optical anemometers where various light sources and methods of modulation of the emitted waves are used.

The Doppler effect is also used to measure the speed of sound in acoustic measuring systems, the so-called sodars. With this method, sound waves are emitted into the atmosphere, which are scattered by inhomogeneities of the moving air medium, which always occur in vortex air flows due to turbulence. The power of the reflected signal allows you to determine the intensity of turbulence, and the Doppler shift gives information about the direction of movement of the air medium. However, the sound is very quickly absorbed by the air and therefore this method is applicable only at distances of no more than 1 km.

In everyday life, one almost never needs to accurately measure the wind speed. However, the Beaufort scale can be used to estimate the wind situation approximately, without using any instruments. In this scale, the speed and strength of the wind are estimated in points, and their value is determined solely with the help of vision. Visually, one ascertains the effect of air flows on surrounding objects or the water surface and, depending on these observations, the wind is evaluated in points from 0 to 12. This scale was developed by Admiral Francis Beaufort of the English Navy and has generally become recognised as effective since 1874 CE. The number of gradations in this scale has varied from 14 to 17. It was not until 1963 CE that the World Meteorological Organization adopted a uniform Beaufort scale which used 12 points.

## 2.12 "Musical Hammer" (Pertaining to the Physics of Sound)

The sonorous part of this toy hammer is a small piece of a plastic tube muffled at one end and closed with a thin lattice at the other. Inside the tube, a small piston with a longitudinal hole can move freely; it is blocked by a thin elastic membrane which begins to vibrate, making a loud squeaking sound, if even a small gust pushes through the piston's hole. To make the hammer squeak, simply strike it against a surface as one might with a typical hammer: the piston, due to its inertia, will move inside the tube and compress the air in the muffled part of the tube. Compressed air passing through the hole in the piston causes the membrane to vibrate and the toy emits its characteristic squeaking sounds in rhythm with the striking. The tonality or frequency of sounds produced by the hammer is determined by the flexibility of the membrane, and their volume or power is determined by the sharpness of the hammer strokes.

The range of propagation of sound waves is determined by the possibility of absorbing the energy of the oscillations of the medium through which the sound propagates. The higher the degree of elasticity of the medium and the fewer obstacles the sound encounters on its way, the

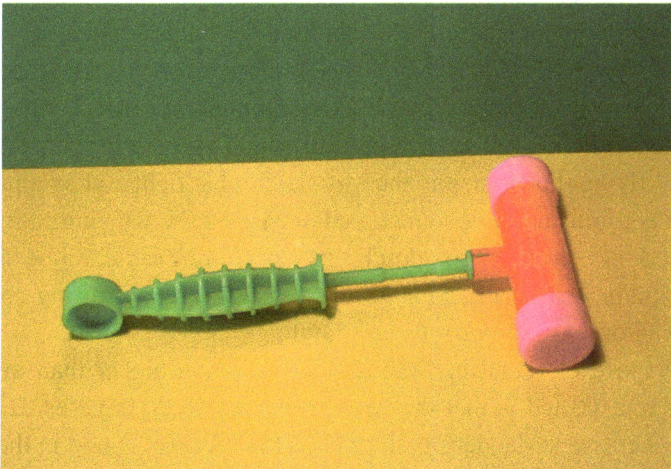

**Figure 2.12**

greater the distance the sound wave front overcomes. In the same environment and at the same power of emission, a sound practically without being absorbed overcomes a distance inversely proportional to its frequency: the higher the frequency, the stronger the absorption. At a sound frequency of 10,000 Hz, its absorption in the air is about 100 times stronger than sound with a frequency of 1,000 Hz. An example of this is the sound of a symphonic orchestra playing at different distances. At a small distance from the orchestra, viols, flutes, and other high-pitch instruments (with a sonic frequency up to about 20,000 Hz) allow the orchestra to express complex harmonies. At a distance from the orchestra, one can only hear the booming sounds of percussion, the frequency of which can reach as low as 20 Hz.

It is interesting that, unlike the propagation range, the speed of propagation of sound waves in any medium practically does not depend on the frequency. It is determined mainly by the aggregate state of this medium: being whether it is a gas, liquid, or solid, and to a somewhat lesser extent on the temperature of said medium. For example, in the air, the speed of sound of almost any frequency, with a temperature difference from −30°C to 120°C, can change by almost 15%, from 300°m/s to 350 m/s. This means that when the air temperature increases by 1°C, the sound velocity increases by about 60 cm/s. The speed of sound in the air varies under different weather conditions and atmospheric pressure, most strongly determined by the height above ocean level for human concerns.

In jet aircraft, usually flying at speeds close to, or exceeding the speed of sound, the speed is measured by *Mach numbers*, which is the ratio of the speed of the aircraft to the speed of sound at the altitude where it is currently flying. For example, the Mig-31 combat fighter at an altitude of 10 km has a cruising supersonic speed of Mach 2.35, which means that at the speed of sound, 1,062 km/h, its speed reaches $2.35 \times 1062 = 2495.7$ km/h. A Boeing 777 passenger aircraft has a cruising speed of Mach 0.84 or $0.84 \times 1062 = 892.08$ km/h.

The speed of sound in various liquid is much higher than in gases. What is unexpected is that the speed of sound in these gases decreases with an increase in the density. However, the speed of sound in these liquids increases with an increase in density. This is demonstrated in Table 2.1.

**Table 2.1**

| Gas | Density g / cm$^3$ | Speed of sound m / s | Liquid | Density g / cm$^3$ | Speed of sound m / s |
|---|---|---|---|---|---|
| Hydrogen | 0.0000899 | 1286 | Gasoline | 0.71 | 1170 |
| Helium | 0.00017846 | 970 | Water | 1 | 1480 |
| Nitrogen | 0.001251 | 334 | Glycerin | 1.26 | 1923 |

According to Table 2.1, sound travels about four and a half times faster in an aquatic environment than in air. In 1827 CE, the speed of sound in water was measured for the first-time with the help of two boats on Lake Geneva, which were 14 km apart. From one of the boats, a bell was lowered into the water by a rope, the impact of which was synchronised with a flash of gunpowder on board. In the second boat there was an observer who, having recorded the moment of the flash, measured the time interval from this flash to the moment when the vibrations from the tolling of the bell were registered by a special tube lowered into the water. The speed of sound measured by them was 1.435 m/s.

Later it was found that with even greater intensity than in the air, the speed of sound in water depends on the temperature of the water. An increase in temperature of one degree Celsius entails an increase in the speed of sound by 3.58 m/s, which is six times faster than in the air. Although water is practically a non-compressible liquid, the speed of sound largely depends on the depth at which measurements are made. At shallow depths, the speed of sound in water is equal to 1.440 m/s. When diving to a depth of 2,000 m below sea level, the speed of sound will be equal to 1.484 m/s. When diving to a depth of 4,000 m below sea level, the speed of sound will be equal to 1.529 m/s. When diving at a depth of 8,000 m below sea level, the speed of sound is estimated to be approximately 1.600 m/s.

As the depth of our world's ocean increases, the water temperature decreases, but at a depth of about 1,200 m below sea level, the temperature becomes constant. It is at this depth that the minimum speed of sound is noted. As one continues deeper into the oceans, with increasing pressure and at a constant temperature, the speed of sound again increases. Since sound waves always tend to go where their speed is reduced, sound is concentrated in this layer of ocean waters (at that same temperature where temperature becomes constant), and it can propagate here over huge distances without much absorption. Such a layer is termed as the deep sound channel, it was discovered in the middle of the last century independently of each other by scientists of the USSR and the USA. It has been discovered that humpback whales use the underwater sound channel for communication.

These and other features of sound propagation in water bodies are studied by a specialised field of science: *hydroacoustics*. This study plays an important role in oceanological research, fish exploration, marine navigation, and the development of useful resources found in the seabed. In solids, sound waves propagate somewhat differently than in liquids and gases.

In fluid and gaseous media, the molecules are weakly interconnected by their electromagnetic forces. The matter of these media can move relative to each other, and when the sound wave front moves, they transmit energy through collisions, while the molecules do not oscillate but instead undulate. Therefore, in these environments, sound can only move in the direction from the sound source, such that there are only longitudinal sound waves. In solid aggregates, molecules (although they may oscillate) are usually held in place by the forces of intermolecular interaction given by the body of molecules which they find themselves within. This intermolecular interaction has an electrostatic origin. This allows solids to retain their shape, as well as to resonate with sound waves which pass through them in various directions rather than just away from the source. Therefore, sound in solids propagates not only by longitudinal waves, but also by transverse waves: waves of compression and expansion propagate in all directions. At the same time, the speed of sound propagation in solids is much higher than in liquids and gases. However, this is only strictly comparable and true of longitudinal sound waves. Transverse waves move much slower, although both speeds are determined by the elastic properties of the particular solid through which the sound wave travels. For example, in plastic lead, the velocities of transverse and longitudinal sound waves are equal to 700 m/s and 2,400 m/s respectively, while in elastic quartz, these velocities are 3,762 m/s and 5,970 m/s, respectively.

An interesting fact is that our Sun is a continuously operating thermo-nuclear reactor, which, although it is very far away from us, emits between 100 W/m$^2$ and 300 W/m$^2$ of sonic energy. If the outer space between our Earth and our Sun was not a vacuum, instead being filled with air, then one could hear the operation of this reactor. Scientists have determined that the total volume of a thermonuclear reaction from the surface of our Sun would reach our Earth with such intensity that one would hear it similarly as if one were standing near a jetliner taking off, it would be at least 120 dB.

## 2.13 "Popping Pistol" (Pertaining to Sound Perception)

One variety of traditional Russian wooden toys is a type of rustic cap gun. It is composed of a hollow wooden cylinder (as the barrel) with a compressible piston (as the handle, inserted on one side of the barrel) and opposite the handle, the barrel can be tightly closed with a cork. The cork is connected to the cylinder by a thin, strong cord and can be inserted into the barrel so tightly that due to the friction force it is able to withhold an air pressure of about 2 atmospheres to 3 atmospheres; such a force is applied if the piston is pushed into the barrel with some effort. When the piston pushed into the cylinder creates an air pressure greater than the cork can withhold, the cork flies out of the barrel with a loud popping sound and is kept from further flight by its cord. This action can easily be repeated as this popping is nothing more than repeated sound pulses. With the help of such a toy, one can begin to explain the physical nature of sound, particularly exploring one of its parameters: tonality.

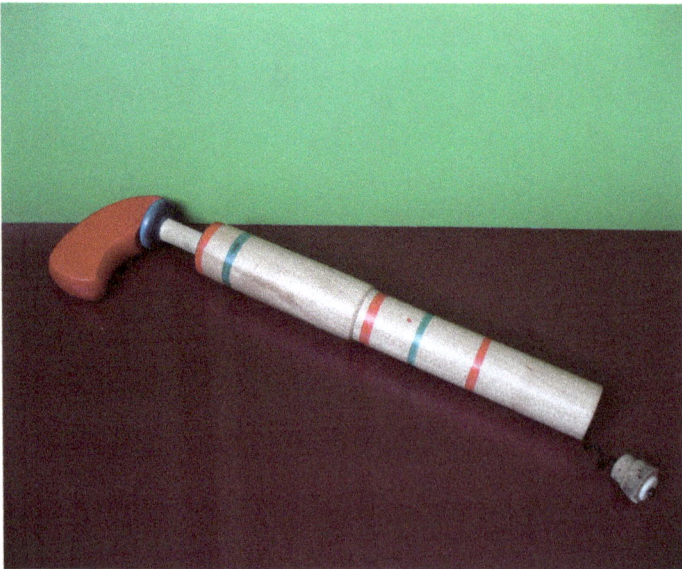

**Figure 2.13**

The study of sound is called *acoustics*. Humans typically can perceive waves of compression and expansion of the air environment with the help of a highly specialised organ, the ear. Within the ear there is a thin membrane, the eardrum, which connects to three special auditory bones. These bones are termed as "hammer", "anvil", and "stirrup". Sound vibrations through the eardrums cause the sound bones to vibrate, which stimulate special auditory receptors (through a special fluid located in the complex system of the inner ear), the excitation of which is transmitted along the auditory nerve to a specialised part of the brain.

In everyday life, the main characteristics of sound for us are its frequency (pitch), amplitude (volume), and speed of propagation. To measure the frequency of any vibration, not strictly limited to sound waves, there are special units of measurement called hertz, named after the German physicist Heinrich Hertz. The physical meaning of these units is very simple: the number of vibrations occurring in one second corresponds to the number of hertz. A good example is our heart, which contracts roughly once per second, giving our heart rate a frequency roughly equal to 1 Hz. In one's youth, one can perceive sound vibrations with frequencies commonly varying between 20 Hz and 24,000 Hz, though one's sensitivity typically will diminish as one ages. The reason for this decrease, in addition to natural biological reasons, is largely the extremity of certain sounds experienced throughout one's life, determined by their duration, volume, and pitch.

Sounds with a frequency of less than 16 Hz are termed as infrasound (a naming convention reminiscent of the light of a lower frequency than our eyes can perceive, termed as infrared, as is discussed in the description of the toy "Laser Shooting Range"), it is not perceived by the human ear yet nevertheless it has many of the properties of humanly perceptible sound waves. The sources of infrasound waves can be vibrations that occur during earthquakes, thunderclaps, the operation of cars, and other mechanisms. Many animals can perceive infrasound auditorily or analogously. It has been noticed that often before earthquakes or tsunamis (as well as other such vast natural disasters caused by movements of the Earth's crust), birds, fish, elephants, and a host of other animals begin to behave restlessly, for no apparent reason to us. This is due to their perceiving infrasound precursors, resulting in their rushing about and panicking.

It is known that jellyfish are aware, insofar as it is possible for them to be aware, of sound vibrations with a frequency between 8 Hz and 13 Hz. Such infrasound waves occur in the sea before a storm begins, and jellyfish, sensing the storm approaching, rush away from the coast. Because infrasound is well transmitted by the aquatic environment, hippopotamuses, crocodiles, and cetaceans have developed the ability to transmit and receive sound signals at this frequency and this allows them to expand their communication capabilities. A person, although they cannot hear infrasound, is quite capable of feeling it. Today, the danger of infrasound exposure to humans has been well established. Powerful infrasound waves affect the work of the heart and brain; even some mental disorders can be exacerbated by these sound waves. Fortunately, one almost never encounters powerful infrasound vibrations in everyday life, and there are always means of protection and noise absorption in hazardous industries.

Sounds with a frequency of more than 20,000 Hz are termed as ultrasound, which is a sound vibration of higher frequencies than perceptible to the ear. In nature, high-frequency ultrasound waves are most often found simultaneously with low-frequency infrasound. Earthquakes and thunderstorms are accompanied by sounds of both ranges; the same can be said about the noise of rain or wind. In the animal world, dolphins are champions in their capabilities in using ultrasound. It is proven that with the help of a complex system of signals at sound frequencies in the range from 20,000 Hz to 25,000 Hz, they are able to communicate. These same frequencies also help them navigate in space with the help of echolocation, as well as having numerous possibilities as a tool in hunting (which they use to stun, disorient, and even kill their prey). Bats also fly at night and hunt using ultrasound, while some butterflies that they hunt are also able to perceive ultrasound and, sensing that they are being hunted, leave the vicinity.

In 1883 CE, the English scientist Francis Galton managed to create sound waves of incredibly high frequency, of about 175,000 Hz, for the first time. To do this, they blew hydrogen under high pressure through a small cylindrical whistle with a very sharp edge. In 1880 CE, two brothers, French physicists Pierre Curie and Jacques Curie, discovered the piezoelectric effect: the formation of an electric charge on the surface of a quartz crystal when it is compressed. This led to the discovery of the appearance of mechanical vibrations in the same crystal when an alternating electric potential is applied to two of its surfaces.

Nowadays, ultrasound has found a wide variety of applications in biology, medicine, and engineering. In biology, ultrasound makes it possible to work with cell membranes, acting on DNA, and it allows one to create artificial mutations within the framework of genetic engineering. Ultrasound devices for diagnostics and therapeutic effects, including in oncology, have been developed and are widely used in medicine. Throughout the natural sciences, ultrasound helps with work in the field of materials science, facilitating the processing of hard and brittle materials, as well as finding applications in non-destructive testing, sonar, and many other fields of study.

## 2.14 "Vuvuzela" (Pertaining to Sonic Power)

The vuvuzela, also called the *kudu horn* or *lepatata*, is originally a musical instrument used by various peoples of southern Africa. Previously, such a horn was made from antelope horns and made extremely loud sounds, which served to alert tribesmen about important events, to corral animals on hunting, as well as being an indispensable accompaniment to dancing and singing. However, since 1970 CE for that year's FIFA World Cup, the vuvuzela has become known around the world as an object of a completely different purpose. Due to its significant capacity for volume and their ability to produce a sound of a very specific key, vuvuzelas are gaining extremely high popularity among especially active fans of football and other game sports.

However, their noisiness has led to the fact that the use of vuvuzelas at major competitions is banned in several countries. The volume of sound, even though it is determined by such a physical parameter as the energy of a sound wave, is a subjective characteristic for a person and is determined by our auditory sensations, which everyone experiences individually, before one even considers one's state of health and age. The unit

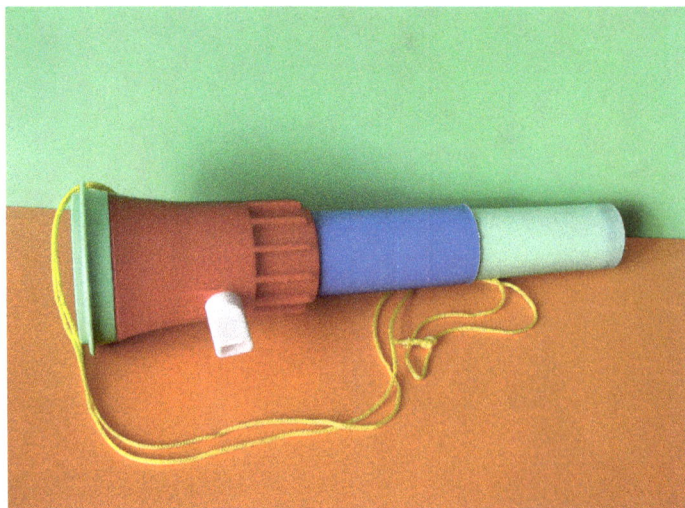

Figure 2.14

of volume in physics are sones, which depend on the frequency of sound and sound pressure.

A sound with a frequency of 1,000 Hz and a value of 1 sone creates a sound wave pressure on our eardrums of about 2 MPa, which corresponds to a pressure of 2 kg/cm$^3$. Other units of sound volume, such as the decibel, are more common and often used in everyday life. The decibel is a special unit of measurement that is not included in the international system of units. It is not a physical quantity, instead it is a mathematical concept defined by a very complex formula. One can compare decibels, although very approximately with percentages. Decibels are also dimensionless and show a comparative sound pressure level. One decibel is a very large value and does not grow linearly with increasing volume. Subjective silence for us: 0 dB does not mean that there is no sound, it only says that the power of the sound wave is so small that our eardrums do not respond to it; 20 dB come to a quiet whisper; 45 dB amount to a loud conversation; 100 dB could be a rock concert; and 160 dB extremely dangerous for a person, corresponding to the volume of a pistol shot beside the ear — this almost inevitably will result in serious damage to one's hearing, accompanied by a rupture of the eardrums. Nowadays, it is considered that background volume between 20 dB and 30 dB is practically harmless for a person, and the maximum permissible limit should be 80 dB.

The propagation of sound around us depends on many factors. Sound waves will propagate differently in various climates and conditions: whether one finds oneself in a forest or inside a building, with atmospheric aridity or humidity, depending on the pitch, even with the same volume! In an open space, volume diminishes in direct proportion to the distance from the source. This is to say that were the distance to be twice as large, the volume would be half as loud. But in rooms and where there are obstacles, there is the factor of sound reflecting off of walls and other such obstacles, leading to multiple secondary sound waves. In large rooms, such as concert halls, the addition of repeatedly reflected sound waves creates reverberation, which provides a sense that one is encompassed by sound. The time of complete attenuation of the reflected sound before its complete disappearance is called the reverberation period. The larger the room, the stronger the reverberation and the more interesting the sound of musical instruments. Sometimes architects attempt to strengthen the reverberation effect with the help of a special arrangement of the walls and ceiling. But sometimes reverberation is also undesirable, a good example being when one is an announcer or working within recording studios.

Similar to reverberation in its physical essence is the echo, which is a reflected sound depending on the speed of propagation of sound waves, as well as on the distance to the reflecting surface.

## 2.15 "Psaltery" (Pertaining to Sound Sources)

The psaltery is a traditional plucked musical instrument of the peoples of Eastern and Northern Europe with similar principles of operation to the harp. This instrument bears a great resemblance to the even more ancient kithara and has been known in Russia since the second century CE. Ancient harps had vein strings with a total number of up to six dozen; modern harps usually have no more than nine strings which are not made from viscera.

All sounds require an oscillating or undulating body causing vibrations of the surrounding air. As such, vocal cords, speakers, and tuning forks (really anything sonorous) can serve as such. Almost all musical instruments contain an oscillating element in their composition or refine the oscillation of the body applied to the instrument. Cymbals and drum membranes vibrate when struck in percussion; strings vibrate in keyboards and guitars when struck; similarly, strings vibrate in viols when a bow is applied to them; vibrating wooden and metal plates emit sound in reeded musical instruments and flutes. Wind instruments that generate sound are generated when the sound moves with inhomogeneities in the instrument, the vibration comes from the body of the player, namely

Figure 2.15

the lips pressed against a mouthpiece. There are also electronic musical instruments that create sound using a membrane oscillated by an electromagnet. To create a coherent sound of a strictly specified frequency, so-called sound or phonon lasers are used.

In the description of the toy "Firecracker", it has already been mentioned that the main characteristic of any wave is its frequency, being the number of vibrations per unit of time, which is measured in hertz. The great scientist of antiquity Pythagoras not only laid the foundations of philosophy as a science and formed the basis of modern mathematics and geometry, they also studied mechanics, cosmology, medicine, art theory, and many other areas of human knowledge. Around 520 BCE, they invented a single-stringed musical instrument known today as a monochord. In this instrument, the stretched string was divided into two parts by a movable support. Two simultaneous pinches could make both parts sound simultaneously. Each part sounded in its own key; the frequency depended on the length of the part. Experimenting with a monochord, Pythagoras discovered that pleasant-to-hear consonances are produced by two simultaneously vibrating parts of one string only when their lengths are related as integers of the first four, which is to say in ratios such as 1:2, 2:3, 3:4 et cetera. This discovery formed the basis of modern music theory with its main musical intervals: the octave (2:1), the quart (4:3), the fifth (3:2), and the triad (3:4:6). Nowadays, such pleasant-sounding consonances are called chords, and chords devoid of musical harmony are called dissonances. Legends also attribute to Pythagoras the invention of pegs, a gear mechanism for adjusting the tension of strings in musical instruments with several parallel strings similar to the lyre or harp.

In all musical instruments, except for the vibrating element that generates sound, there is a special cavity that increases the volume of sound, promoting resonance. Sound resonance is a sharp increase in the amplitude of vibrations of the air, or any other medium through which sound waves travels, when the frequency of the external sound generator coincides with the frequency of the already vibrating medium. To ensure a stable sound resonance in the air environment, it is necessary to have a closed volume of air. This condition will provide repeated reflection from the walls to the already oscillating waves, and the external source will provide a decrease in energy during absorption.

In guitars, violins, and other string instruments, the chamber for resonance is a hollow wooden body. In wind instruments, the cavities of pipes
made of a variety of materials, including but not limited to copper, reed,
or wood, play the role of this space for resonance. The shape of this cavity
plays a very important role, as it is this space which provides the local
maximum amount of reflection of sound waves of a particular frequency.
There is a simple explanation for this: reflection is optimised if the
reflected sound wave has a length that fits into the distance between the
reflecting surfaces a number of times which would closely aligns with an
integer. The sound frequency is inversely proportional to the sound wave-
length: $\lambda = \frac{V}{v}$, where $\lambda$ is the sound wave length, $v$ is the sound frequency,
and $V$ is the speed of sound. The choice of the resonating material also
depends on the prevailing frequency range of a particular musical instrument. This is explained by the fact that the intensity of sonic wave absorption is determined by the elasticity of the resonating material and the
roughness of the absorbing surface.

In life, often engineers are more concerned with facilitating sound
insulation rather than resonance. In almost all industries, in construction
and in transport, there are high levels of noise exposure, from which the
hearing of workers requires special protection. Nowadays, in large cities,
the problem of noise insulation of our housing is extremely urgent. To
solve these problems, acoustic scientists have developed methods of noise
reduction based on changing the direction of noise radiation, incorporat-
ing rational architecture and the layout of buildings, as well as the strate-
gic use of sound-absorbing materials and coatings.

# Chapter 3

# Electricity

## 3.1 "Van de Graaf Generator" (Pertaining to Electrostatics)

This toy has many parts concealed within its plastic handle: two AA batteries, an electric motor, an endless conveyor made of silk ribbon with strips of metal foil glued on it, a current collector that removes charges from metal strips, and an electric switch button. At the upper end of the handle there is a metal contact, which, connecting to the current collector inside the handle, simultaneously comes into close contact with the inner surface of the cardboard cylinder slotted onto the toy.

If one were to take the assembled toy, examine it in detail, including activating it by pressing the button located on the handle, it would be very difficult to understand how to play with it and for what purposes it is intended without special instructions and additional explanations. By pressing the button, you can only hear the soft buzzing of the motor within; nothing more will happen. However, if you put a strip of very thin dacron film covered with an even thinner layer of metal and place it atop the very edge of the cardboard tube, then, after briefly running the motor, the cut petals of the film will begin to lift themselves and the shapeless strip will turn into a three-dimensional structure. Moreover, even if afterwards you were to stop pressing the button, halting the motor, and then were you to shake off the structure of the metallised film from the cardboard tube, it will hover over it, as if weightless. A butterfly figurine cut out of the same metallised film, which is also included in the toy set, will

**Figure 3.1**

also behave in an equally unusual way. Being placed on the edge of the cardboard tube, after a few seconds of the motor running, the butterfly will suddenly break away from the tube and will hover over it even when the motor is turned off.

This unusual behaviour from the metallised film is explained in the manifestation of the forces of electrostatic repulsion. No one knows exactly what electricity is to this day, although people have been acquainted with it since ancient times. These first encounters, back in the sixth century BCE, pointed out the property of rubbed amber to attract light dust particles. Contemporary to this, the unusual properties of electric fish were described and ways of using them for medicinal purposes were proposed. Thus, a few decades into the Common Era, Pedanius Dioscorides used electric eel strokes to treat gout and chronic headaches. In addition, ancient Roman authors mention the glow of the tops of masts on ships and the pointed parts of tall buildings before a thunderstorm. Later this phenomenon was called the Saint Elmo's fire, after the Catholic patron saint of sailors. Within the works of Lucretius, who lived in the first century BCE, one can find a type of scientific concept of magnetic properties, according to which a magnet emits the smallest particles, "grains" that push the air apart, forming voids where iron atoms move in to fill the vacated space.

During the Middle Ages, for reasons that are not entirely clear, the interest of scientists in the manifestations of electrical or magnetic properties did not manifest itself in any way. Only since the beginning of the forties of the eighteenth century CE has there been a real outbreak of interest in electricity, not only among scientists, but also in the broad strata of educated society. It is noteworthy that this interest did not have any practical purpose, rather being motivated by curiosity and wonderment.

In this time of the first investigations of electrical and magnetic properties in the Modern Age, the most common theory of electricity suggested that it was the flow of a certain weightless liquid from one body to another, forming an electric current, and the presence or absence of a reserve of such a liquid provided an electrostatic charge. Such famous scientists as Benjamin Franklin and Alessandro Volta, Charles Coulomb, and Simeon Poisson were supporters of this point of view. It is interesting that although by this time a possible connection between electricity and magnetism was already being viewed, they could not come up with any experiments to test it at that time. This was done only at the beginning of the nineteenth century CE.

Another direction in the theory of electricity in the middle of the eighteenth century CE was the principle of *long-range action*. According to this theory, the carriers of electric and magnetic interaction are weightless and elusive aether-particles. The founder of such ideas was Rene Descartes; these ideas were also adhered to by Lomonosov Mikhail Vasil'jevich and Leonard Euler.

The introduction of the concept of "field" into science by the English scientist Michael Faraday followed by the creation of the theory of the electromagnetic field by James Maxwell brought about the next stage in the development of the theory of electricity. Further development of Maxwell's theory, which became classical electrodynamics at the end of the nineteenth century CE, led to the formation of the foundations of quantum field theory by Max Planck at the beginning of the twentieth century CE, then subsequently to the creation of the special theory of relativity by Albert Einstein.

All stages of the development of the theory of electricity were accompanied by the development of the art of experiment and the improvement of practical research, first pertaining to electrostatics and magnetism, and then to electrodynamics. Perhaps the first experiments that became known after the ancient experiments of Thales of Miletus with rubbed amber were the experiments of the English doctor and natural scientist William Gilbert, also with amber and various magnets.

In 1672 CE, the German physicist Otto von Guericke described a special mechanism that rotated a sphere with a diameter of 15 m comprising cast sulphur, capable of carrying out strong electrification. The device which they designed may have been the very first electric generator. The English physicist Francis Hawksbee perfected the Guericke machine — they no longer had a solid ball of sulphur rotating and rubbing their hands, but a hollow ball of glass with air pumped out of it. In their electricity generator, they observed an unusual glow — today this phenomenon would be termed as ionisation. Such experiments were demonstrated by Hawksbee at meetings of the Royal Society for the Development of Natural Sciences in London. Isaac Newton, who was its honorary president at that time, showed considerable interest in them.

In 1745 CE, in the city of Leyden, Pieter van Musschenbroeck discovered the possibility of accumulating a significant electric charge in a glass phial filled with mercury. Thus, the first volumetric capacitor was invented, later called the *Leyden jar*, which served as a powerful instigation to improve electric machines.

In 1766 CE, the English optician Jesse Ramsden built an electric machine in which a rotating glass disc was rubbed with two pads covered with *an amalgam* (an alloy of mercury with other metals), such that an electric charge was accumulated using several Leyden jars.

The largest electrostatic friction machine ever built was an electrostatic generator built in 1784 CE by John Cuthbertson in Harlem, made from two glass discs with a diameter of 1.65 m and four sets of Leyden jars. This generator could be charged to a potential of 33,000 V.

In 1754 CE, John Canton, using an electroscope improved by themself, discovered that two touching, lightweight metal balls begin to repel when a third charged ball of an external charge approaches them. Thus, electrostatic induction was discovered. With the help of this discovery, a method was found for obtaining a charge on conductive bodies without touching an external charge and without generating charge through rubbing; it became known as "electrification through influence". In this method, a third ball of a certain electric charge is brought to two touching metal balls that do not have an electric charge. When within the electro-static field of the charged ball, free electrons are redistributed on the surface of the touching balls; some of these electrons will be attracted to or repelled by the charge, creating a field, depending on the nature of the charge. Following this field interaction, the touching balls are disconnected from each other, and the charged ball is removed, such that on each of the disconnected balls there is a charge of the opposite sign from that of the introduced third ball.

This is the method used by Alessandro Volta when creating the world's first *electrophore*, being an electricity generator based on induction. Volta's electrophore is arranged very simply: there are two discs which are aligned untouching one above the other. One is a smooth disc made of a mixture of resin, wax, and turpentine cast in a metal mandrel. Above that disc is a disc made of metal. Where the design is simple, the logic of its operation is not. Firstly, the metal mandrel is connected to the ground with a metal wire, thus grounding it. Secondly, the metal plate is removed from the disc of sulphur and wax, rubbing the disc with fur, thus electrifying and creating a negative charge on its surface. Thirdly, a metal disc is placed on the surface of a charged non-conductive disc. In this case, the metal disc should not come into contact with the metal mandrel. Positive charges on the resin disc will attract negative charges on the metal surfaces of the mandrel and the disc closest to it, whilst positive charges from the mandrel will pass through the metal wire into the ground, leaving the negative charges to remain on the upper sheet. Since the charge on the resin plate has practically not changed (it does not conduct an electric current), such a cycle can be repeated many times and each time a positively charged metal sheet can be obtained without a new rubbing of the resin disc, solely due to the redistribution of charges on the metal sheet and the ground.

The cyclic process of obtaining charges was first automated by the English inventor Francis Ronalds. Using the weight of a pendulum driven by a clock mechanism, a metal plate would periodically connect to an electrified insulator. Ronalds used a similar mechanism when creating the world's first electric telegraph.

In 1788, William Nicholson proposed moving a metal plate by rotating it over a charged insulator, thereby creating the first prototype of modern electrophoretic machines.

Electrophoretic machines with two rotating glass discs, close to the modern form, were designed almost simultaneously by two German physicists. These contemporary claimants to being the original inventor were August Toepler and Wilhelm Holtz. Holtz's machine was a little simpler in design, but it worked more efficiently than Toepler's.

In 1883 CE, the English inventor James Wimshurst proposed an even more advanced version of these machines. Its design was comprised of two rotating glass discs, two Leyden jars, and two electrodes in the form of steel balls. Such a device is a very visual demonstration of electrostatics; it is still widely used in school physics laboratories.

The end of the nineteenth century CE was marked by an unprecedented rapid growth of large-scale research in the field of the structure of matter. As a necessary tool, devices that enabled the generation of electric charges with vast possible potential were required, potentials in the order of at least hundreds of thousands of volts, ideally tens to hundreds of millions of volts. Electrophoretic machines of the time were not capable of providing this output. In 1933 CE, the American physicist Robert Van de Graaf of Princeton University created an electric generator that could reach a voltage of approximately 7,000,000 V. The operational principle of the Van de Graaf generator is the same as that of electrophoretic machines, through electrification by influence. The main difference was that in their new generator, small charges obtained by electrifying the battery were transferred to the main storage device by an endless conveyor belt made of insulating material, on which flexible metal strips of foil were glued, rather than via a glass disc.

A very ingenious solution in the Van de Graaf generator was to transfer single charges from the foil strips to the inner surface of the battery. Since the storage device was a thin-walled, hollow metal sphere with rounded edges, any new single charge from the inner surface, under the influence of the Coulomb force, moved to the outer surface. Even with a very high potential on the outer surface of the sphere, its potential always remained zero inside; new small single charges fell into the storage device without interference, immediately transferring to the surface. It is this property of hollow spherical electrostatic charge accumulators that allowed them to obtain such high voltage.

Despite meagre efficiency, not exceeding 20%, Van de Graaf generators were instrumental in the acceleration of charged particles for nuclear research in the first half of the twentieth century CE, as well as in materials science and in the development of high voltage technologies. Such generators are still used today in individual scientific centres, although they are much more often found in modern scientific museums and in large educational centres, where they work as an extremely effective demonstration device.

## 3.2 "Radio-controlled Toy Car" (Pertaining to Radio Communication)

Radio-controlled toys appeared in the second half of the last century, quite recently when compared with some of the toys which were explored in the first chapter. The principle of radio communication is based on the joint operation of two electronic devices, one being an electromagnetic wave generator, the other being their receiver, which can be used when at a distance from each other. The simplest source of electromagnetic waves are electrical discharges that are repeated at a certain frequency. In 1887 CE, Heinrich Hertz discovered that if an open metal frame, which has small balls suspended in the cavity within the bounds of the frame, is located near an electric circuit with such discharges, then discharges also appear between those suspended balls, slightly weakened and almost simultaneous.

The signal required for the operation of the toy is transmitted through the air using electromagnetic waves, the frequency and length of which is determined by the time between discharges. The wavelength $\lambda$ depends upon the time period $T$ between its two peaks, as well as the velocity $V$ of

Figure 3.2

the propagation of the wave front for ordinary oscillatory systems, as is described by the traditional formula: $\lambda = VT$. Since the duration of one oscillation (the value of $T$) and the frequency of the wave $v$ are inversely proportional to $T = \frac{1}{V}$, and the speed of propagation of electromagnetic waves is equal to the speed of light ($V = C$), it can be noted that the frequency of electromagnetic waves depends on their length as: $\lambda = \frac{C}{v}$.

Heinrich Hertz invented a special device that created repeated, regular, rapid discharges, thus generating electromagnetic waves. This device would come to be named the *Hertz vibrator* in honour of them. Nowadays, it is more commonly known as an *open oscillatory circuit*. At the time of Hertz, an open metal frame was used as a receiver of electromagnetic waves, likely as this is the simplest *resonator*. Based on the spark generators designed by Popov Aljeksandr Stjepanovich and Guglielmo Marconi at the very end of the nineteenth century CE, the first radio transmitters and radio receivers were built. These technologies would be crucial for the development of practical radiotelegraphic radio communication. These early long-distance communication technologies employed code analogous to alphabets. A famous example would be *Morse code*, a system of notation where each letter corresponded to its own combination of dots and dashes, and the letters from these were formed into words and phrases. The radio transmitters transmitted short and long pulses of electromagnetic waves corresponding to these dots and dashes. In the electrical circuit of the radio receiver, there was a special device called *a coherer*, which changed its electrical conductivity under the influence of electromagnetic waves. The coherer controlled an electromagnetic coil with a spring-loaded anchor, which was attracted and released depending on the frequency of the received electromagnetic waves. The writing node was connected to the anchor, allowing it to print on a paper ribbon a chain of dots and dashes exactly coinciding with the one that was transmitted by the radio transmitter.

The first coherers were a glass tube with metallic or coal filings, in which sparks jumped under the influence of electromagnetic waves. The electrical conductivity would increase, and it would become possible to register an electromagnetic wave pulse. Another design of the coherer was a glass tube with slightly oxidised iron electrodes under voltage and a drop of mercury between them. When exposed to radio waves, the oxide film on the electrodes would radiate at a certain frequency, causing the electrical resistance in the circuit to drop sharply, thus registering the radio transmission. Such a coherer was first used by Marconi for communication between ships of the Italian Navy at the very end of the nineteenth century CE.

The frequencies of the first radio-transmitting devices were low, which did not allow transmission of radio waves with an acceptable quality corresponding to the variation in sonic frequency of human speech. However, the rapid development of electrical engineering led to the appearance of radio transmitters, where alternating electric current and an electric arc were used instead of a spark. Such transmitters could already work at frequencies up to 50,000 Hz, allowing human speech to be transmitted quite well by way of radio waves. Soon electronic generators would appear, working first on vacuum tubes, and then on semiconductors. Compact, miniature, and economical radio transmitters were emerging that operated in a vast frequency range of radio waves. They have come to find application in almost all spheres of human activity: radio broadcasting, radio communications and radar.

All radio waves are in the frequency range of 30 kHz to 3,000 GHz, divided into the following sub-bands:

- SDV: ultra-long wavelengths spanning 10 km to 10,000 km, with frequencies between 30 Hz and 30 kHz;
- DV: long wavelengths spanning 1 km to 10 km, with frequencies between 30 kHz and 300 kHz;
- SV: medium wavelengths spanning 100 m to 1 km, with frequencies between 300 kHz and 3 MHz;
- KV: short wavelengths spanning 10 m to 100 m, with frequencies between 3 MHz and 30 MHz;
- SKV: ultra-short wavelengths spanning 10 mm to 10 m, with frequencies between 30 MHz and 30 GHz;
- MMV: millimetre wavelengths spanning 1 mm to 10 mm, with frequencies between 30 GHz and 300 GHz;
- SMMV: submillimetre wavelengths spanning 100 $\mu$m to 1 mm, with frequencies between 300 GHz and 3,000 GHz.

Depending on the frequencies of radio waves, their penetrating ability is different, so they propagate in the atmosphere in different ways. Consequently, each of the types of waves is used in different ways.

Ultra-long radio waves can penetrate underground and through water, allowing them to be used for communication with submarines and underground structures.

Long radio waves are well suited to bypass obstacles and can spread over our Earth's surface for thousands of kilometres, even circumnavigating the globe. They are used for radiotelegraphic communication, radio broadcasting, and long-distance radio navigation.

Medium radio waves traverse the irregularities of our Earth's surface worse and are absorbed more by our Earth than long radio waves, but they can be reflected from the ionosphere, especially well at night, and are also able to bend around our Earth's surface. Thus, this range is the most common for radio broadcasting.

Short radio waves are even more strongly absorbed by our Earth's surface and spread only in a straight line and over short distances due to the absorption of their energy by the planet's surface. They are also reflected from the ionosphere regardless of the time of day. They can go around the planet consuming energy for their generation much less than for waves with a lower frequency. They are mainly used for radio broadcasting and radio communications.

Ultrashort radio waves can (mostly) only propagate in a straight line and are completely absorbed by any irregularities of our Earth's surface. But for this range, it is possible to create narrowly directed antennae, which allows you to concentrate their radiation. Another feature of ultrashort radio waves is that they do not reflect from the high ionised layers of our Earth's atmosphere but are able to reflect from the lowest layer, being the troposphere. This feature allows one to employ tropospheric radio communication, which allows one to transmit information simultaneously through several channels.

Ultra-short radio waves, whose length ranges from 1 m to 10 m, form the metre range, which are typically used for radio communications, television, and radio navigation over relatively short distances. Ultra-short radio waves, whose length ranges from 1 m to 100 mm, form the decimetre range; they are used in various types of broadcasting (radio communications, radar, television, satellite, and tropospheric radio communications) as well as by cell phones. Ultra-short radio waves, whose length ranges from 10 mm to 100 mm, form the centimetre range; they are used in satellite communication, television systems, tropospheric radio communications, wireless computer networks, radar, microwave ovens, as well as for radio control of various models and drones.

Until recently, radio waves of the millimetre range, whose length ranges from 1 mm to 10 mm, did not have a wide scope of application, since there was not a sufficiently developed base of electronic components to create appropriate radio transceiver equipment. Over the past ten years, significant progress has been made in this area, and the range has become actively used for narrow-range radio communication at short distances, radar, radio astronomy, and medicine. It may be that this becomes the dominant range of information transmission for new generations of interconnected computer networks.

Submillimetre radio waves, whose length ranges from 100 $\mu$m to 1 mm, are an intermediate link between infrared and ultrahigh-frequency electromagnetic radiation. Currently, this range is not used for the purpose of transmitting information, but it is recognised as promising and large-scale research is being conducted in this direction. Such low-energy radio waves are safe for humans, so they are widely used in medicine when scanning human organs, as well as for baggage control at airports.

Any transmission of information by radio waves requires variations in certain aspects of the electromagnetic spectrum, principally changes of frequency, amplitude, or phase. These changes are referred to as modulations. These modulations can be compared to the behaviour of waves at the beach to help one to understand their properties. Frequency is comparable to the time between the cresting of waves, such that the modulation of frequency is comparable to waves cresting at a faster or slower rate (the shorter the time between crests, the shorter that one would imagine a wavelength to be). Amplitude is comparable to the height of the waves as they roll into the shore, such that the modulation of amplitude is comparable to waves having deeper troughs between their peaks (the deeper that the waves are, the greater that one would imagine the magnitude — which is to say the brightness — of the radio waves to be). Phase is less intuitive to describe, though it more closely reflects in its nature the metaphor of the behaviour of waves at the beach than the image of waves of consistent rate of cresting or height. Phase modulation is when there is a change in duration between waves — it is comparable to if a cresting wave was followed by another wave which did not leave a full trough between them, even if both waves had the same swell. Frequency modulation is a special form of phase modulation which has a full trough but changes the distance between crests.

An example of the application of frequency modulation would be in sound transmission for television broadcasting. An example of the application of amplitude modulation would be in radio broadcasting on long, medium, and short waves, as well as in television broadcasting. An example of the application of phase modulation is in radio communications in digital radio broadcasting and television broadcasting.

Transceiver radio devices for the radio control of models and toys commonly employ both frequency modulation and amplitude modulation, often for radio waves between 54.4 MHz and 70 MHz, though ranges between 2.4 GHz and 5.8 GHz are also known to be used. Frequency modulation is more often employed when the distance from the transmitter to the model is small. Frequency modulation is typically cheaper than employment of amplitude modulation, though this discrepancy is paltry. Radio-controlled models using amplitude modulation are more resistant to interference, making them more suitable for long-distance control.

## 3.3 "Helicopter with Automatic Controls" (Pertaining to Automatic Control Systems)

Until recently these toys were not common despite the logic of the technology being well known, as the components for their construction were prohibitively heavy for good duration of utility. Nowadays, lithium-ion batteries' capacity and neodymium electric motors' powerfulness, considering their relative lightness, have made it possible to create this flying toy.

The remote control for toy helicopters typically only requires one hand to operate, though this may vary between brands. The toy in Figure 3.3, once activated, will begin to spin its blades, resulting in the aircraft levitating to a height of approximately a meter and a half or so. Without subsequent command, this toy will start to descend steadily, dropping to approximately 10 cm to 15 cm above the surface below. This surface can be any horizontal surface realistically, be that the floor, the table, or one's own palm. Once settled, the aircraft will take off again to a similar height as prior. This cycling of ascent and descent can last as long as 10 minutes, ceasing at the depletion of the battery.

Figure 3.3

This automatic movement is explained by the device which the helicopter employs: there is an infrared distance sensor in the bottom of the helicopter. Such a sensor consists of a light-emitting diode or chip (commonly termed as LED) that emits electromagnetic radiation with a maximum of approximately 940 nm (much smaller than submillimetre radio waves), a receiver of the reflected radiation in the form of a photodiode of the same wavelength as the emitting body, and an electronic control circuit. The toy, with a fascinating mode of flight, is a clear example of an automatic device with negative feedback, which will be described below.

The word "automatic "comes from the Ancient Greek words meaning "self-acting". The first automatic devices which were invented by humanity likely were snares and traps, designed for facilitating the easy hunt of birds and animals. These devices, while certainly not complicated, were real automatons. Skilfully installed and disguised, these devices, without any further operation by the hunter, when exposed to game on its triggering mechanisms, would be triggered automatically. A description of more advanced automatic devices can be found in the works of the ancient scientists Heron of Alexandria and Vitruvius.

An example of such a device would be the temple automaton of ancient Egypt, dating back to about the second century BCE. When dropping a copper coin into a special slot, the automaton would pour a portion of blessed water for ablution to the person who paid. Another example is the automatic doors, again in a temple, which would open by themselves when a fire was lit in the sacrificial hearth. In a metal chamber concealed beneath the altar, fire heated the air, expanding to push water through a siphon into a huge bucket. Said bucket was suspended on chains in a system of pulleys and counterweights that opened the doors as the bucket became heavier. After the fire was extinguished, the air in the ball cooled, decreasing in volume. Subsequently, the water would be sucked into the siphon, the empty bucket would rise, and the doors would solemnly close.

These ancient automatic devices operated themselves according to their precise mechanisms. The cycles of the automata could only be repeated; the operations themselves did not depend upon further operations of the internal parts of the mechanism. In modern technology, such mechanisms are called *automation system with open protocol communication*. Other *automatic feedback devices* differ in that they can change the duration, sequence, and total set of operations performed automatically, depending on the operation of their own parts, and considering external conditions. It is according to this algorithm that the toy helicopter described above flies, tracking its distance to any horizontal surface.

The first automatic feedback machines might have been the water clock, one of which was created in the second century BCE by the Ancient Greek scientist Ctesibius. In their clock, the water pouring out in a thin stream from a large, elevated vessel provided the energy for the rotation of a cylinder. This cylinder was segmented into divisions and numbers indicating the time of day as the pouring water filled another vessel. In this second vessel, one half of a curved syphon was lowered to its very bottom. As the second vessel was filled, the tube of the syphon was filled with water. When the water level rose to the upper bend of the siphon and filled it to completion, then, in accordance with the law of communicating vessels, all the water from the second vessel would drain through the descending length of the syphon into a third vessel. Concurrently, the draining water would rotate the blades of a small turbine, which through a set of gears forced the hand of a figure resembling cupid to move. This figure would show that the next day had come by gesturing with its finger. Finally, a door in the bottom of the third vessel would activate, emptying the third vessel and preparing the mechanism to work again.

The logic of water clocks, also known as clepsydrae, regularly work in the plumbing equipment of modern homes. Surprisingly, in the ancient world, automatic mechanisms rarely were used for domestic purposes, instead being used only for religious purposes or entertainment. Even during the Middle Ages and the Renaissance, more advanced mechanical automata, of which countless examples can be imagined varying from clockwork dolls to music boxes to tower clocks, continued to serve only the purposes of entertainment.

The advent of the steam engine in the eighteenth century CE led to a surge in the development and production of automatic machines. At first production was done by small studios. Soon large industrial enterprises would emerge, the technical equipment of which was based on equipment with varying degrees of automation. Following closely on steam engines' proverbial heels, the mass development and introduction of electric automatic devices with input and feedback systems began, all of which would lead to the first mechanical computers being created.

Among the ingenious automatic devices of that time, one might note the centrifugal speed controller of the steam engine shaft. This tool was invented in 1784 CE by the English mechanic James Watt, for the purpose of maintaining a constant speed of rotation. It was arranged as such: two spring-loaded caps were symmetrically attached to hinged levers which were themselves attached on a vertical axis connected by gears to the main rotating axis of the steam engine. The hinged levers, being preloaded by springs, kept the pressure relief valve in the steam boiler closed. When a certain rotation speed was exceeded, the centrifugal force of inertia acting on the rotating loads overcame the elastic force of the springs. Subsequently, the caps on the hinged levers were removed from the axis, simultaneously opening the steam pressure relief valve. When the steam pressure decreased, the rotation speed would decrease. Subsequently, the caps would return to their original position, closing the valve, allowing the pressure to increase again. The automatic adjustment cycle thus was repeated, operating in a negative feedback loop.

Similarly of interest were the ways to adjust the temperature in poultry egg incubators without using electricity, which were in use at the end of the nineteenth century CE. The heat source in such incubators was a kerosene lamp located in the lower part of the incubator. The automatic regulator was a sealed, lentil-shaped capsule soldered from copper foil and filled with a solution of esters that boiled at a temperature between 32°C and 40°C. The hot air from the lamp required to maintain the incubating temperature passed through the incubator through a thin-walled iron pipe, the upper part of which was covered with a light lid on a lever with a hinge that rested on the capsule. When the temperature in the incubator exceeded the required temperature for incubation, a simple chain of events would generate the force necessary to operate the lid. The solution in the capsule would boil such that the pressure of its vapours forced the capsule to expand, thus the lever connected to it would ease the lid open. Through the opened pipe, hot air could bypass the incubation chamber, allowing the capsule to cool, thus decreasing in volume, and subsequently closing the lid. This cycle could then repeat itself automatically.

The most profound implementation of the principles of automation was found not in mechanics, but first in electrical engineering, and then in electronics. Initially, the main element of automation in any electrical device was an electric relay: a switch that closes or opens one electrical circuit, depending on the presence of current in another electrical circuit. In 1831 CE, the first electric relay was invented by the American scientist and electrical engineer Joseph Henry. It was an electromagnetic coil with a spring-loaded steel core, which, in the presence of current in the coil, closed the contacts of an external electrical circuit. Similar relays are widely distributed to this day in electrical circuits through which electric currents of significant magnitude flow.

By 1880 CE, the next significant step in the development of automation of electrical devices transpired after years of research by numerous scientists. John Swan, taking refinements made by Thomas Edison, produced the first lamp diodes, which were called detectors at that time. Today it is one of the main elements of electrical circuits, the main feature of which is the ability to pass current only in one direction and lock it in the other. A little later, in 1899 CE, the German physicist Karl Braun managed to create a solid-state diode. In 1907 CE, the American inventor Lee de Forest received a patent for another fundamentally new electronic device, the lamp triode. Their device was a glass flask filled with magnetically neutral gas, (later it was removed from designs), into which three electrodes were inserted. If one passed a small electric current between two interconnected thin metal grids, then a cloud of electrons escaped from the grid, which made it possible to pass a much larger current between the third electrode and one of the first two electrodes. The triode allowed one not only to pass both direct and alternating current in a single direction, but to amplify it as well.

The designs of lamp triodes were further refined as time went on. Until the late forties of the twentieth century CE, these triodes, along with solid-state and paw diodes, formed the basis of all electronic automatic devices. However, in 1947 CE, the Americans William Shockley, John Bardeen and Walter Bright managed to create the world's first semiconductor triode. Such elements of electronics began to be called transistors and for their invention in 1956 CE, all three inventors were awarded the Nobel Prize in Physics. The invention of the first transistor marked the beginning of a new stage in the development of electronic automation using semiconductor devices based on crystalline silicon and germanium. The dimensions of the new electronic devices became much smaller than the lamp ones, which made it possible to create miniature and reliable electronic devices and led to another renaissance, bringing electronic computing technology to the world. The previously existing computers based on electronic paws and relays were extremely cumbersome and slow, inconvenient, and expensive. Inevitably, this hindered the development of their scope of application, inhibiting the improvement of programming languages. The appearance of transistors made it possible to achieve unprecedented computing speeds and create a sufficiently large computational memory capacity, which in turn opened the way for the development of high-level programming languages.

In the second half of the last century, with the advent of nanotechnology, transistors, diodes, capacitors, resistors, and all other components of electronics became miniscule, making them able to be combined into integrated circuits. Nowadays, working with desktops or pocket computers has become common place all over the world. The algorithms of software have become sophisticated, narrowly focused, and universal programming languages have appeared. Global information networks (such as the Internet) have been created, but all this rests on the tried and true algorithm of input and feedback.

## 3.4 "Beetle on a Solar Battery" (Pertaining to Photovoltaics)

This toy is very diminutive, being all of 3 cm without its antennae. It has a charming way of moving on a flat and brightly illumined surface, devoid of wheels or other traditional propellers, but with the help of thin spring-loaded wire legs. This gait comes from a very small electric motor with a weight on the shaft, which is offset by a millimetre from the axis of rotation. When the weight rotates, a centrifugal force of inertia is created, acting through the legs on the surface of the table, resulting in a reaction force pushing the legs away from the table. The front legs of the beetle are slightly shorter than the hind legs, so the sum of the reaction forces acting on two pairs of legs creates a resulting force vector directed upward and forward. The energy for the rotation of the electric motor is provided by a small solar battery on the beetle's back.

In 1839 CE, the prerequisites for the creation of solar cells with the discovery by Alexandre Becquerel of the photoelectric effect phenomenon. Following this discovery, experimenting with chemical current sources, they noticed that if two electrodes (in this case, of gold and platinum) were placed in a vessel with two different electrolytes separated by

Figure 3.4

a semi-permeable partition, then the electric current generated by them in the light was greater than in the dark. At the same time, it was found that the current strength increased with a change in the spectrum of incident light from red to green yet decreased as it moved towards violet from green. However, the efficiency of using solar energy in a chemical current source with additional lighting did not exceed 1%. Such meagre efficiency precluded further development of this for the purposes of electrical engineering.

After the theoretical explanation of the photoelectric effect by Albert Einstein in 1905 CE, the possibility of creating photovoltaic sources of electricity with a sufficiently high efficiency was proven possible, but technical progress in this direction did not happen for almost half a century. A radical change occurred only after silicon-based photovoltaic cells with an efficiency of about four percent were invented in 1954 CE. Before that, work was underway on selenium-based cells with an efficiency of less than 3%.

The principle of operation of silicon solar cells is based on the peculiarity of silicon to form two different phases of electron excitation when enriched with a small amount of different trace elements. In one of them, for example, with arsenic additives, silicon atoms on the outer electron shells have several electrons with weak binding energy and an excess of free electrons may exist in such a material. This type is called an *N* semiconductor, the letter *N* owes its origin to the word negative. This refers to physical concept that an excess of electrons leads to the formation of a negative electric charge. In another variety, for example, with indium additives, all the electrons have a high binding energy, and there are free places for "foreign" electrons in the outer orbit. This type is called a *P* semiconductor, the symbol *P* owes its origin to the word positive. This follows from a lack of electrons being responsible for the positive charge. Under certain conditions, electrons from one type of semiconductor can pass into a semiconductor of another type. Such a transition is commonly called a *P N semiconductor transition*. This model is the basis for creating most modern solar cells.

Nowadays, there are many types of solar cells with different efficiency coefficients. This is the result of properties of the crystallisation of the silicon atoms. The highest efficiency, approaching 22%, can be achieved in the most refined single-crystal silicon-based solar cells. They are considered the most durable and high-quality; they have a blue appearance of a darker hue compared to the rest. Solar panels with polycrystalline silicon are made from many smaller crystals, which are easier and currently cheaper to produce. They can provide an efficiency of up to 18%, their colour is slightly lighter and brighter. It has been suggested that these batteries work quite well in diffuse light. Amorphous silicon can be used in solar cells made from a thin film. These cells have a light grey appearance; the efficiency of these cells is still relatively low, only reaching 10%. Although today approximately five-sixths of all used solar cells are monocrystalline, experts predict great prospects for film solar cells based on improvements in technologies related to amorphous silicon.

# Chapter 4

# Optics

## 4.1 "Laser Shooting Range" (Pertaining to the Electromagnetic Nature of Light)

Nowadays, lasers have lost some of their shock factors, becoming quite a common part of many toys; however, they remain interesting in their operation. An example of a toy using a laser would be a toy gun with a target. The gun comprises a plastic body, batteries contained within the handle, a speaker within the barrel, an electronic control circuit and the red laser in the tip aligned with the barrel. The target comprises a plastic body, also containing a control circuit, speaker, and batteries, but this contains a photocell, which acts as a receiver for the red laser's light. When the photocell reacts to the laser shining, it will play a tune through its own speaker. The speaker in the gun reacts to the trigger being pulled with a firing sound. While the toy is simple as a game, it is much more advanced in its composition than tin cans lined up to be shot at with a slingshot or a cap gun (as was the way to play this game in the childhood of today's adults, when a laser toy still had the ability to surprise). The laser is of particular interest for the purposes of this scientific exploration.

Light has long demanded investigation, leading to many perplexing discoveries. Aristotle suggested that "light is something emitted from the eyes", though this suggestion has aged worse than many of their other ideas. Modern scientific endeavours have helped illumine the physical properties of light, though the metaphysical properties remain a mystery. In the seventeenth century CE, Willebrod Snellius (known to many English-speaking

**Figure 4.1**

people as Snell) wrote on the laws of refraction. Similarly, the algebraic geometry proposed by their contemporary Rene Descartes allowed a codification of the relations between light and the changes to its path as it passes through different mediums (as described in Snell's law).

In 1665 CE, the Jesuit Francesco Maria Grimaldi published their treatise *Physicomathesis de Lumine, Coloribus et Iride Aliisque Adnexis*, within which they described the phenomenon of diffraction. Using camera obscura, they were able to study the behaviour of shadows and determine that light did not move simply in a rectilinear fashion but had variations in intensity. They even drew comparison to how waves interfere with themselves as they ripple over each other much as light interferes with itself, which in the age of quantum physics shows their prescience.

In 1690 CE, Christiaan Huyghens suggested within their treatise *Traite de la Lumière* that the wave-like properties of light could be accounted for by the medium through which light moved, naming this elastic medium the Aether.

Isaac Newton's conceptual approach to light was characterised by the same material discretion and finitude as their anticipations of movement of bodies in relation to each other. They imagined light as very small particles and explained this wave-like interference property as inter-particular inter-action. Their studies of light spanned 40 years of intermittent research from 1664 CE, laying the foundations and embedding the conceits of particularity for the modern study of light. Newton had considered that light had both the properties of a wave and a particle: within their work *Opticks*, they state that light is "nothing but pressure or motion propagated through the aether". Newton coined the term *spectrum* when referring to the disambiguation of light into its constitutive frequencies when passed through a prism. While the rainbow has long mystified humanity, its properties being known since Ancient Rome and its significance being marked in the oldest biblical stories, it was Newton who first described the linearity of the spectrum. Their ideas of the spectrum of light are released to the public in their adage, 35 years after they conducted their first experiments into refraction, reflection, and the decomposition of sunlight. Newton also further refined the design of the telescope from Galileo Galilei's designs, moving from *concentrating lens* to minute *reflective lens*.

The investigation of the properties of light was taken further by Augustin Fresnel, who investigated the interference of *polarised light*. Where ordinary light has transverse vibrations in all directions, polarised light has transverse vibrations that are aligned along certain axes. Where two beams of light which are polarised in parallel planes will always form an interference pattern, two beams which are polarised in perpendicular planes will never form an interference pattern. To account for these physical properties, Fresnel suggested that light represented transverse waves through the aether, permeating all material bodies.

In 1848 CE, Michael Faraday discovered the ability of a magnetic field to interact with polarised light by rotating the plane of polarisation. In 1865 CE, James Maxwell, thanks to their theory of electromagnetism, was able to calculate the speed of propagation of electromagnetic waves: it was equal to the speed of light measured at that time. Maxwell, considering the investigations of Faraday, concluded that the propagation of electromagnetic waves coincided with that of light as these waves were of the same physical nature, differing only in wavelength. Heinrich Hertz experimentally confirmed the existence of electromagnetic waves, finding that their speed of propagation was equal to that of light. The nature of light as a wave through aether was accepted as indisputable.

The subsequent development of the theory of light came from further investigations of phenomena pertaining to diffraction and interference. In 1800 CE, William Herschel discovered ultraviolet radiation. In 1801 CE, Johann Ritter discovered infrared light. Hindsight would reveal that they were simply outside the field of visible light, but each lay adjacent to either side on the spectrum (as their names explain). Visible light is a very specific part of the spectrum, which is roughly characterised as follows:

(from wavelengths longer than ultraviolet light)

| Color | Lowest wavelength (m) | Longest wavelength (m) |
|---|---|---|
| Purple | $3.8 \times 10^{-7}$ | $4.4 \times 10^{-7}$ |
| Deep Blue | $4.4 \times 10^{-7}$ | $4.85 \times 10^{-7}$ |
| Bright Blue | $4.85 \times 10^{-7}$ | $5.0 \times 10^{-7}$ |
| Green | $5.0 \times 10^{-7}$ | $5.65 \times 10^{-7}$ |
| Yellow | $5.65 \times 10^{-7}$ | $5.9 \times 10^{-7}$ |
| Orange | $5.9 \times 10^{-7}$ | $6.25 \times 10^{-7}$ |
| Red | $6.25 \times 10^{-7}$ | $7.4 \times 10^{-7}$ |

(to wavelengths shorter than infrared light)

Different parts of the spectrum can have totally different functions materially, resulting from differences in penetration and interference. High-frequency waves are often used in medicine, such as with radiography, and low-frequency waves are often used in common tools and toys, such as microwave ovens and radio transmission.

By the end of the nineteenth century CE, numerous experiments were conducted with lamps made to emit light through electric excitation of various gases. During experimentation, it was discovered that the light emitted during this excitation ended out being from particular parts of the spectrum rather than from all across the spectrum, as was thought to be the case from burning objects. It was later discovered that individual atoms of any chemical element would have predictable emissions during excitation to the point of emission of light. In 1890 CE, Johannes Rydberg proposed a mathematical formula describing the wavelengths for a broader spectrum of electromagnetic radiation for hydrogen beyond the visible range. According to this formula, the wavelength of the radiation obeys a simple rule: $\frac{1}{\lambda} = R\left(\frac{1}{n^2} - \frac{1}{k^2}\right)$, where $n$ and $k$ are natural numbers and $R$ is a numerical constant named after Rydberg. According to this formula, for $n = 1$, a change in the value of $k$ from 2 to infinity gives a series of lines in the ultraviolet part of the spectrum (these are called the Lyman series). For $n = 2$, a change in $k$ from 3 to infinity gives a series of lines in the visible part of the spectrum (these are called the Balmer series). For $n = 3$, a change in the value of $k$ from 4 to infinity gives a series of lines in the infrared part of the spectrum (these are called the Paschen series).

This was a discovery based purely on empiricism, theory would have to play catch up. In fact, it took a little more than 20 years for theory to catch up. Niels Bohr proposed a new model for the structure of the hydrogen atom, which would lead them to their ideas of an entirely new metaphysical underpinning to physics, being quantum mechanics.

At the end of the nineteenth century CE, the wave theory of light was shaken by new experimental data. The essence of the paradox was that within the wave theory, the power of electromagnetic radiation should increase as the wavelength decreases, tending towards infinity. Numerous experiments were not compatible with this understanding. The radiation spectrum of black bodies (bodies which are in thermal equilibrium with their environment and absorb all electromagnetic radiation upon them) was one experiment to refute the wave theory. Hertz's discovery of *the photoelectric effect* in 1887 CE was also inexplicable: why would electrons be emitted by certain illumined materials?

In 1900 CE, Max Planck managed to solve the problem of black body radiation. Their suggestion was that light was not emitted in a continuous stream, but in portions (which they referred to as *quanta*). They also determined that the energy of such quanta was dependent on the frequency of the wavelength, as described by the following elegant formula: $E_0 = h\nu$, where $\nu$ is the frequency of the oscillations of the wave and $h$ is a specific constant later named for Planck. Since the frequency of electromagnetic radiation $\nu$ is related to the speed of light $C$ and the wavelength $\lambda$ by the ratio $\nu = \frac{C}{\lambda}$, the energy of a quantum of light, as proposed by Plank could also be written as $E_0 = h\frac{C}{\lambda}$. The exact value of the Planck constant has been repeatedly redefined, and nowadays it is recorded as

$6.626070 \times 10^{-34}$ m$^2$ kg/s. Planck's hypothesis, in which light can behave as both a quantised wave or/and a particle, aided Albert Einstein in their explanation of the photoelectric effect in 1905 CE. They suggested that overcoming the forces that hold an electron in a specific shell of a metal atom is possible due to the additional energy imparted by a quantum of light.

In 1913 CE, using the idea of quantisation of electron energy levels, as alluded to previously, Niels Bohr built a new model of the structure of the hydrogen atom. According to Bohr's theory, each atom has strict energy inputs and outputs when there is excitation of the atom relating to their stationary energy levels. In these states, atoms do not emit energy. Stationary energy states can only be changed if there is energy as an input to the atom or an output from the atom, manifested through the absorption or emission of a quantum of energy, being a photon. As an example, if the energy state changed by way of the energy decreasing, then light would be emitted. The difference in energy is described by the following formula: $E_2 - E_1 = h\nu$, where $E_1$ and $E_2$ are the atomic energies before and after, $h$ is Planck's constant and $\nu$ is the frequency of radiation. When a photon is absorbed, the energy level of the atom increases and the photon itself disappears.

In the model of the hydrogen atom proposed by Bohr, the energy level of the atom is determined mainly by the energy of an electron that rotates around a nucleus (in the case of hydrogen, consisting of a single proton, thus there is only one electron to consider). In this case, the energy of the electron depends on its distance from the nucleus. There is hypothetically no limit to the orbits which an electron might occupy, but they are all quantised. This is to say that an electron, no matter what orbit it is in, must have a strictly defined energy. Bohr determined the dependence of this energy on the electron's orbit by the following formula: $mvRi = nh$, where $m$ is the mass of the electron, $v$ is the orbital velocity, $r_n$ is the radius of the $n$th orbit, $n$ is the orbital shell, and $h$ is Planck's constant. The number $n$ is called the main quantum number and can take any positive integer towards infinity as its value. For $n = 1$, the electron is in the orbital shell closest to the nucleus, its rotational energy is at its highest and its orbital velocity is $v \approx \frac{c}{137}$. This state is often referred to as the ground state. Higher orbits (with higher $n$ values) have more energy, the speed of electrons increases as the orbit becomes more distant from the nucleus.

Thus, Bohr proposed a theory based on the idea of stationary energy states of the atom, describing a new mechanism of radiation, which would become known as *the first text describing quantum mechanics.* This theory did not end out holding water as it considered more massive atoms.

Quantum mechanics continued to develop. In 1923 CE, Louis de Broglie suggested that the wave nature of quanta of electromagnetic radiation was a property of all particles of matter. De Broglie proposed a mathematical formulation of this idea, which described the equivalent relationship between wavelength, mass, and speed. At the end of 1924 CE, Wolfgang Pauli formulated their famous *prohibition principle.* This was the rule of electron shell distribution, which allowed one to build a theoretical model of the electron shells of an atom and played a fundamental role in the formation of a new quantum mechanics. At about the same time, Werner Heisenberg, abandoning the figurative idea of electrons rotating around the nucleus and proceeding only from the observable parameters of the atomic microcosm, developed a tabular method for calculating the energy levels of electrons. Heisenberg continued to develop this method and, working together with Max Born and Ernst Pascual Jordan, published a fundamentally new version of quantum theory called *matrix mechanics* in 1925 CE.

Matrix mechanics raised the ideas of the structure of the Boron atom to a new level, solving the problem of mathematically describing the transition of electrons between any energy levels. The main disadvantage of matrix mechanics was the absence of any visibility of the physical phenomena. In 1926 CE, Erwin Schrodinger, the conceptual inventor of *wave quantum mechanics,* managed to eliminate this drawback to a certain extent. This tool of thought, using differential equations instead of matrix calculations, was able to describe the quantum world in terms close to those of classical physics and turned out to be less abstracted than Heisenberg's mechanics. Schrodinger's mechanics are based on the concept of a wave function, which, although not an observable quantity, allows one to describe the waveform at a given time with the help of mathematics. For example, for an electron, the Schrodinger wave function shows how large the electric charge density is at a certain point in time at a certain point in space. At the same time, the electron itself in

Schrodinger's view can in no way be a particle, it is a kind of three-dimensional wave packet. A rough one-dimensional analogy of such a package can be imagined in the form of a segment of a wave launched by a hand strike on a stretched rope.

The matrix formulation of Heisenberg's quantum mechanics and the Schrodinger wave formulation are mathematically equivalent: the results of calculations of the same characteristics of the atomic microcosm performed using different methods would be the same. However, a *physical interpretation* of what was being defined figuratively had yet to be developed at the time of these models' development.

In 1927 CE, Heisenberg discovered *the uncertainty principle*, stating that it is impossible for quantum particles to simultaneously measure the location and momentum or energy and time with an accuracy exceeding a certain value containing Planck's constant. Mathematically, this principle is written as $\Delta p \Delta q \geq \frac{h}{2\pi}$, where $\Delta p$ and $\Delta q$ are inaccuracies in the determination of the momentum and coordinate, $h$ is Planck's constant, and $\pi$ has its classical value. That same year, Niels Bohr discovered *the complementarity principle*, according to which the particle and wave properties are complementary to each other, but mutually exclusive. In one or another experiment, either the wave or particle nature of a material particle can manifest itself, but not at the same time, while a complete quantum mechanical description of such a particle is given only by a combination of wave and corpuscular properties.

Both Heisenberg's uncertainty principle and Bohr's complementarity principle only partially helped in the physical interpretation of the quantum mechanical picture of the world. The most common *is the Copenhagen interpretation*, which is based on Bohr's statement that quantum mechanics is a completely complete theory, and quantum reality is determined solely by the process of measurement or observation. Moreover, measuring and observation devices are inextricably linked with the object under study, and it is impossible to separate them. This means that, for example, an electron cannot exist at a given point in the real world if its parameters are not measured or a simple observation is not carried out. This view was supported by Heisenberg but was questioned by Schrodinger and Einstein. Although the Copenhagen interpretation is generally accepted today, there is still no complete agreement in understanding the physical meaning of quantum mechanics in the scientific world, and there are more than a dozen different interpretations of it.

Nowadays, quantum mechanics is one of the most fruitful and experimentally confirmed theories in modern science. Its main areas of application are nuclear research, nuclear energy, chemistry, electronics, and modern technologies operating on the subatomic scale. One of the most important practical achievements of quantum mechanics was the invention of lasers. Lasers are sources of powerful electromagnetic radiation in a very narrow wavelength range. This specificity accounts for the feature called monochromatic electromagnetic radiation, as the laser appears to be a particular colour to people when in the visible part of the spectrum, such as the laser which this toy employs.

In 1917 CE, Albert Einstein anticipated that atoms in an excited state can emit induced light under the influence of another light incident on them. Nowadays, it is known that the operation of any laser is because an atom in an excited state, when interacting with an external photon incident on it, can forcibly emit another photon, and since the incident photon does not disappear, the light flux increases. An indispensable condition for such amplification is the equality of the energy of the emitted photon and the energy difference of all levels of the atom before and after the stimulated radiation. An extremely valuable feature of such stimulated radiation is that it is coherent, which is to say that it completely coincides in frequency, phase, and plane of polarisation with the same characteristics of the photon that caused this radiation.

In 1960 CE, Theodore Meiman built the first quantum generator of electromagnetic radiation in the visible range. In this pulsed laser, the atoms of a ruby crystal were brought into an excited state by the action of photons emitted by a flash lamp. Photons were accumulated using two mirrors (one of which was translucent) located at the ends of the ruby crystal and a coherent light flux pulse with a wavelength of 694.3 nm was emitted through the translucent mirror.

Today, hundreds of different laser systems have been created that differ in the chemical composition and the aggregate state of the working medium, being the substance consisting of excited atoms emitting coherent radiation. Working mediums need not only be made from solid substances: various gas mixtures, liquids and semiconductors can work as mediums. There are schemes of nuclear-pumped lasers and X-ray lasers. Various dielectric crystals, glass, transparent ceramics, and polymers with activating additives are used in solid-state lasers. Gas lasers usually use argon, helium, neon, carbon dioxide, nitrogen, and metal vapours as constitutive of the working medium. Liquid lasers work on dye solutions as well as liquids with ions of rare earth metals and other elements. In semiconductor lasers, a whole range of semiconductor substances are used, the most common among these being gallium arsenide. The procedures for feeding the excitable material with external energy, a process commonly called *laser pumping*, are also various and diverse. Optical pumping takes place using a pulsed or constant source of photons, some examples being specialised electric lamps, gas-discharge tubes, or other lasers. Electric pumping is carried out by way of an electric discharge between the electrodes in gas lasers or an electric current flowing in semiconductor lasers. There are also gas-dynamic pumping, chemical pumping, and supplying energy to the body of quantum generators, just to name a few from many.

Nowadays, with the rapid development of nanotechnology, the most common lasers are those that use semiconductor materials as a working medium, occupying an intermediate position between metals and dielectrics. In laser technology, the most common semiconductors are gallium arsenide, aluminium gallium arsenide, gallium phosphide, gallium nitride, and indium gallium nitride. Various chemical compounds of cadmium, selenium, sulphur, zinc, indium, phosphorus, arsenic, and antimony may also be used.

At present, lasers have found their widest application in civil, military and household appliances, as well as in medicine, in addition to being used in scientific research. The most common, due to extensive technological development and relative affordability, are semiconductor lasers. Toys often employ such lasers.

When playing with kittens, laser pointers of various colours from the visible part of the spectrum are used to create an uncatchable point of light. Prices on these lasers vary due to factors in their construction. Green pointers are typically priced slightly higher than others due to a complex optical system containing a crystal of yttrium vanadate combined with a crystal of titanium-potassium phosphate, allowing them to emit in the range of 510–530 nm; they are as close as possible to the sensitivity of the human retina. This feature of green laser radiation requires special care in handling such devices: even with a small radiation power, when it directly hits the retina of the eye, the risk of damage to it is extremely high. Blue pointers appeared on the market not so long ago and therefore are rarer, usually operating at a wavelength of 405–473 nm, having a power of less than 500 MW; their semiconductors comprise gallium, aluminium nitrides, or aluminium-yttrium garnet with neodymium additives. Their beam is close to ultraviolet light; thus, when illumining some objects, it can cause fluorescence, appearing as an afterglow of the surface of these objects. However, pointers based on a red laser diode emitting at a wavelength of 634–670 nm are most common at present. As a semiconductor, these lasers use a chemical compound of arsenic, aluminium, and gallium. A bright red beam is emitted by lasers of even low power, as little as from 1 MW, but usually the power of laser pointers is about 5 MW. A similar laser is used in the toy for demonstration.

## 4.2 "Kaleidoscope" (Pertaining to Geometric Optics)

*The kaleidoscope* is a simple device. It appears to be like a telescope if one were to look upon it rather than to use it, but the purposes of its operation could hardly be more dissimilar. Instead of a system of lenses, numerous mirror strips are spread out along the longitudinal axis of the kaleido-scope, facing each other with reflective surfaces. Beyond the mirror strips, as one travels down the kaleidoscope away from the eye, a small compart-ment is fenced off with two round glasses. The glass farthest from the eyepiece is frosted to prevent interference (and precluding any telescopic functionality), the glass closest to the eyepiece is transparent. Various small effects are between these glasses: things such as coloured glass, foil or beads. If you look through the eyepiece and rotate the kaleidoscope, the interspersed effluvia will, repeatedly re-reflecting inside the prism, create a continuous series of non-repeating, geometrically correct, symmetrical, and some might think very beautiful patterns.

There are many variations of the design of kaleidoscopes. One of the varieties, the *athaneidoscope*, has no compartment with rolling objects. Rotating the kaleidoscope tube in one's hands, one looks through a prism

**Figure 4.2**

at stationary sets of coloured stones, beads, chains and even insects, spread out and illuminated on a social table or on an ordinary table. There are kaleidoscopes with incorporated lighting sources. In liquid kaleidoscopes, instead of rolling coloured glasses, a mixture of several colour liquids that are insoluble in each other is poured; sometimes metal sequins, beads, or other objects are added to the liquids.

In 1816 CE, Scottish physicist David Brewster invented and patented the first kaleidoscope. Being interested in optical research and experimenting with the reflection of light rays from various surfaces, they found a pattern linking the degree of polarisation, the angle of incidence of the light beam, and the transparency of the dielectric. This regularity was later called *Brewster's law*, which is described by the following formula: $tg\alpha_B = n$, where $\alpha_B$ is *Brewster's angle*, and $n$ is *the index of optical refraction*. The index of optical refraction is equal to the ratio of the speed of light in a vacuum to the speed of light in the medium under consideration, which in this case would be dielectric. In engineering, the indicator $n$ is referred to as the optical refractive index; the Brewster angle is understood as a special angle of incidence of a light beam characteristic of each dielectric material, at which the degree of polarisation is maximum. You can read about polarisation in the description of the toy "Laser Shooting Range".

Optics is the field of physics that studies the emission, propagation of light and how it interacts with matter. These aspects are studied through three major disciplines:

1. geometric optics;
2. physical optics;
3. physiological optics.

*Geometric optics*, without going into the depths of the questions of the origin and nature of light, considers the laws of reflection, refraction, and propagation of rectilinear light rays in optically homogeneous media. The core concept explored through geometric optics is light as a beam. It can be imagined as a line along which transverse waves of electromagnetic radiation of a very limited frequency propagate with great speed. The behaviour of such rays in geometric optics is described by four basic laws:

1. *The law of straightness*: In a homogeneous medium, light propagates in a straight line.
2. *The law of reflection*: The incident and reflected light rays, as well as those perpendicular to the plane of reflection, lie in the same plane. The angles formed by the incident beam and reflected beam with the reflection plane are always equal.
3. *The law of refraction*: If the incident ray enters a transparent medium of a different density, it will split into two rays: one shall be reflected, remaining in the first medium; the other shall be refracted, entering the medium at a changed angle. In this case, all rays will be in the same plane, whether it is the incident, reflected, or refracted rays. The relations between the angles are as follows:
   • The angles of incidence and reflection are equal to each other.
   • The sine of the angle between the incident beam and the axis perpendicular to the axis of the surface between the media, multiplied by the refractive index of the incident medium, is equal to the same product for the medium in which the beam is refracted. In 1620 CE, Dutch astronomer Willebrord Snellius established this pattern. In 1630 CE, René Descartes proposed its mathematical notation as follows: $n_1 \times \sin \alpha_1 = n_2 \times \sin \alpha_2$, where $n_1$ and $n_2$ refer to the optical properties of the medium of incidence and refraction, $\alpha_1$ and $\alpha_2$ are the angles between the normal to the plane of the discontinuity and rays of incidence and refraction.
4. *The law of independence of light rays*: According to this law, light streams of low intensity do not interact with each other when meeting and crossing.

*Physical optics* differs from geometric optics in that it studies the deep processes of light emission, its nature and internal structure, as well as all physical phenomena associated with light. Within the framework of this science, a number of physical features of light have been experimentally established. Where light manifests itself as a transverse wave, the nature of light propagation in inhomogeneous optically transparent media: diffraction, interference, polarisation, all these phenomena are the subject of consideration of wave optics. But the emission of light, its absorption, the photoelectric effect, spectral characteristics, and general patterns of inter-action with matter indicate that light is an intermittent stream of particles. The description of the behaviour of these quanta of light lies within the framework of quantum optics.

The main subjects of the study of *physiological optics* are the percep-tion of light fluxes by the organs of vision of animals and humans, as well as photochemical processes occurring under the action of light energy on biological objects, bioluminescence (light emission due to the decay of chemicals) and biofluorescence (re-emission of light quanta) and photosynthesis.

# Chapter 5

# Homemade Toys

## 5.1 Glider (Made from plastic film and bending plastic straws)

Developing the theme of a gliding flight, you can make a simple but elegant flying toy out of straws and plastic wrap.

For constructing this toy, you will require:

- Four straws with a diameter of 6 mm with bending corrugations
- A piece of plastic film from any package cut to dimensions of approximately 40 cm by 30 cm
- Electrical insulation tape
- Transparent adhesive tape
- Scissors

Method:

1. Bend two straws in their corrugations and connect them with pieces of insulation tape, as shown in Figure 5.2.
2. Cut one straw and connect it to the other, as shown in Figure 5.3.
3. Insert two connected straws into the gap between the other two straws, as shown in Figure 5.4.
4. Spread out and flatten the plastic film on the table and put the frame of straws on it. Carefully adhere the frame to the film with adhesive

**Figure 5.1**

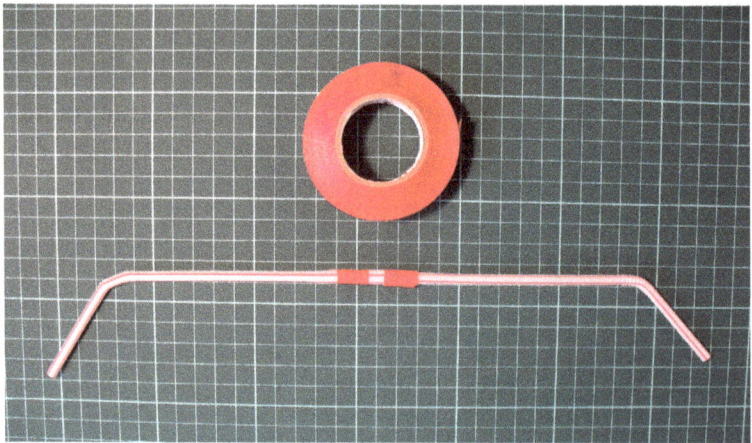

**Figure 5.2**

tape, keeping the overlap of the tape and the film over the straw to less than 5 mm, as shown in Figure 5.5.

5.  Use scissors to cut the film, as shown in Figure 5.6.
6.  Wind insulation tape around the nose of the glider thrice for balancing, as shown in Figure 5.7.

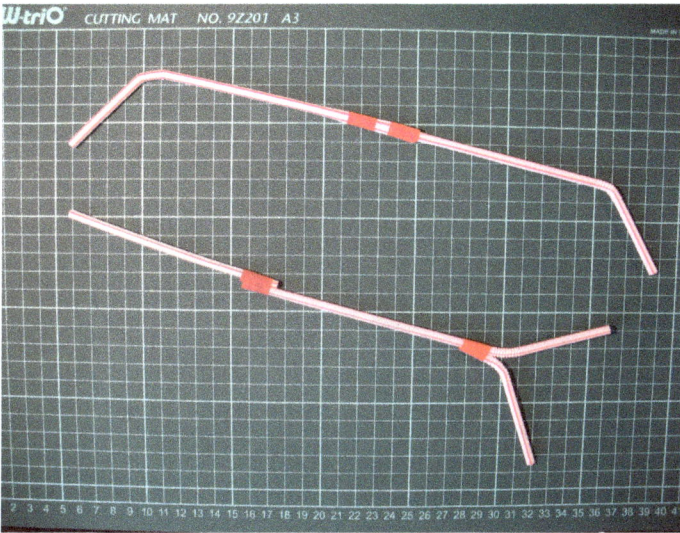

**Figure 5.3**

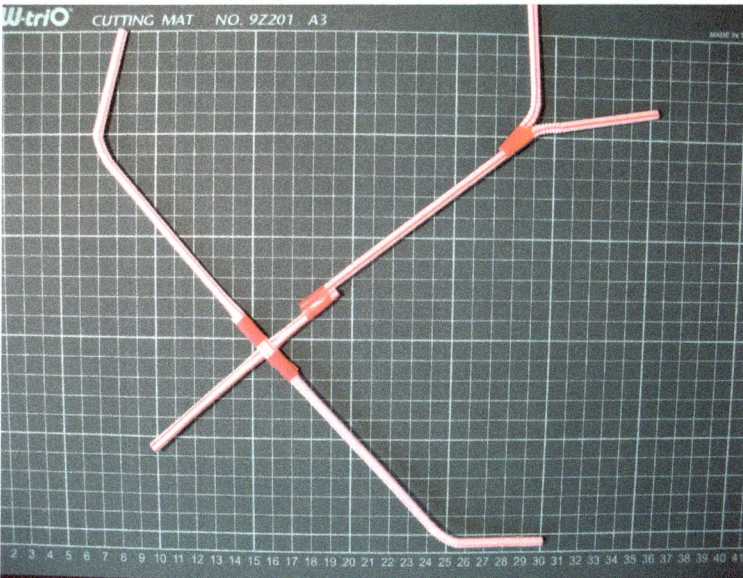

**Figure 5.4**

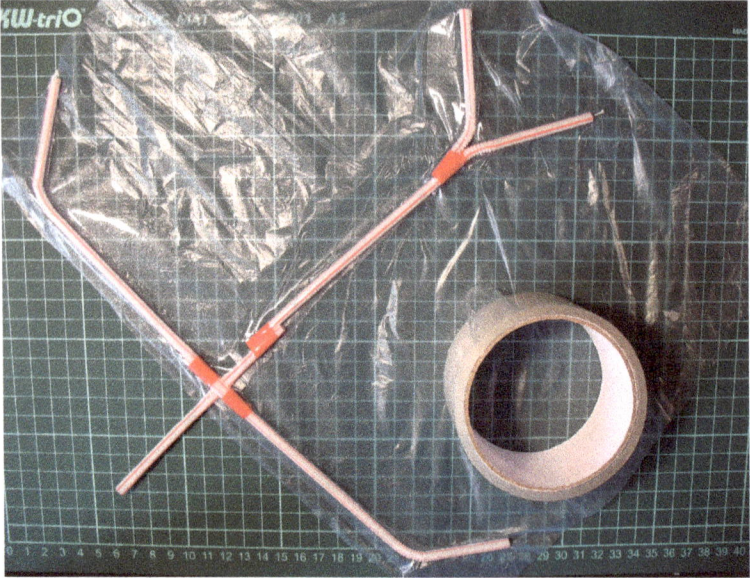

**Figure 5.5**

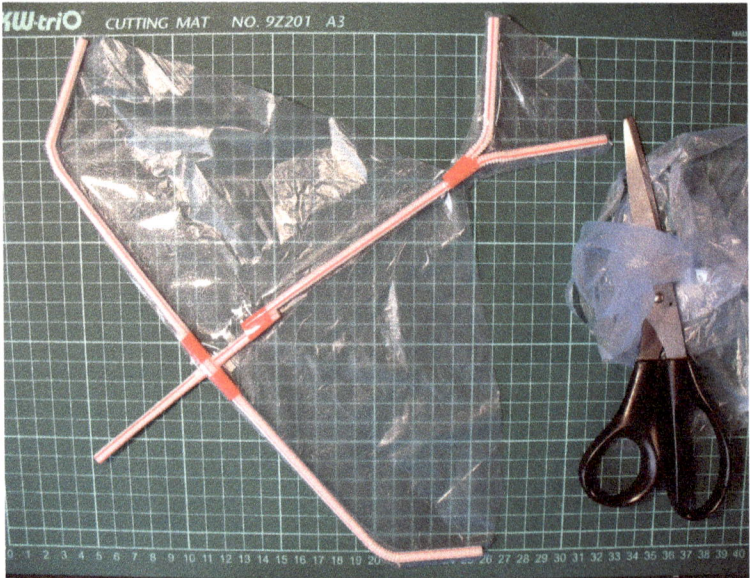

**Figure 5.6**

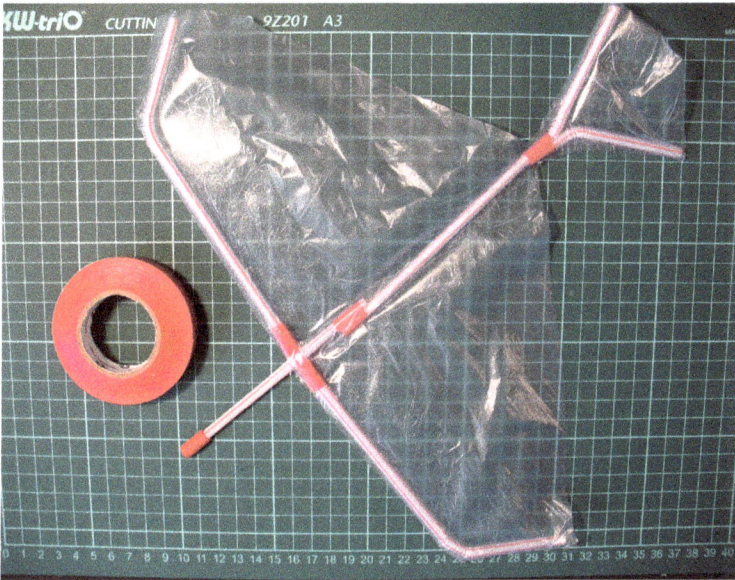

**Figure 5.7**

When launching one's glider, one will find that experimenting with the angle between the wings and the angle of inclination of the tail will alter flight. One may need to unwind or add insulation tape on the nose of the glider to balance it.

## 5.2 Gyroscopic Precession (Displayed using a plastic bottle top)

The simplest precessing gyroscope, which rotates in a sliding bearing on a thread, can be made from a bottle top from a plastic bottle, a piece of a straw and a skewer, as shown in Figure 5.8.

For constructing this toy, you will require:

- Plastic bottle top
- Thread
- Electrical insulation tape
- Bamboo skewer
- A straw of no particular diameter
- Scissors

Method:

1. Make a hole at the centre of the bottle top, as shown in Figure 5.9.

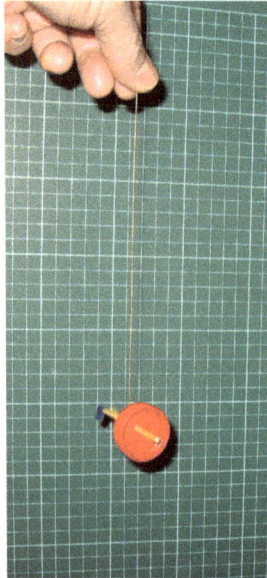

**Figure 5.8**

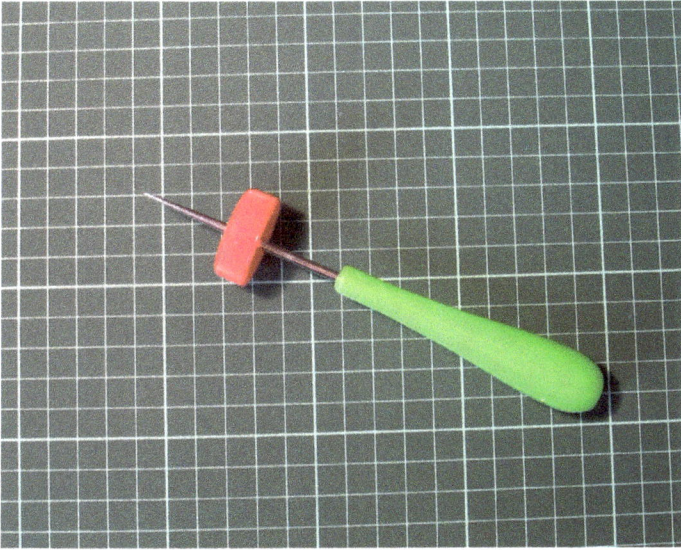

**Figure 5.9**

**Figure 5.10**

2. Insert a piece of a skewer about 5 cm long into the hole, as shown in Figure 5.10.
3. Cut a piece from the straw, as shown in Figure 5.11.

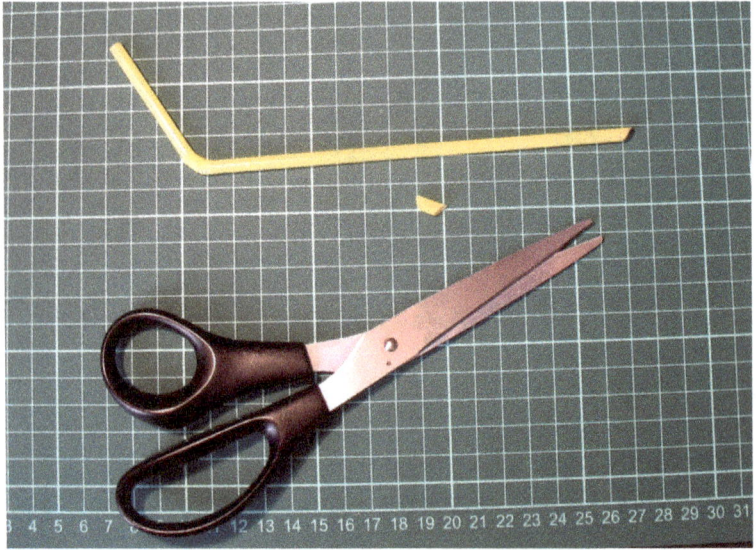

**Figure 5.11**

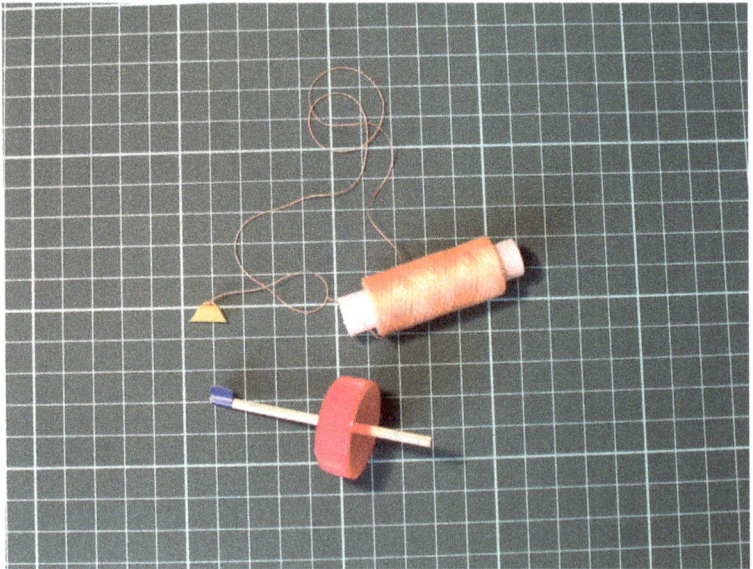

**Figure 5.12**

4.  Tie thread to the piece of straw and adhere a strip of tape to the very edge of the skewer, as shown in Figure 5.12.

Before starting the gyroscope, you need to thread a bamboo skewer into a piece of a straw as shown at the beginning of the description in Figure 5.8. To make the gyroscope precess, you need to hold the gyroscope upright with two fingers by the edge of the skewer without tape, pulling the string as steadily and quickly as one can, then releasing it. With a little training, one can make the gyroscope rotate in two planes for three to five seconds. After that, the friction force will extinguish the inertia of the gyroscope and it will start to rotate vertically, but the piece of tape will not let the toy fall to the floor and roll away, thus, the experiments can be repeated immediately.

## 5.3  Friction Force (Displayed using a cardboard box)

This simple toy is a cardboard box that slides freely down a stretched thread. This toy is interesting for its unusual behaviour — it can be made to hang on the thread for any length of time even though it will slide around freely otherwise. After asking the person playing with the toy to try to explain the behaviour of the toy and receiving, most likely, incorrect answers, the hidden device of the toy can be described exactly. The retention effect comes from the creation of a friction force at the points of contact of the thread with the toy at the time of increased thread tension.

The general view of the toy at the time of its hanging is shown in Figure 5.13.

For constructing this toy, you will require:

- A sheet of A4 coloured craft paper
- Thread

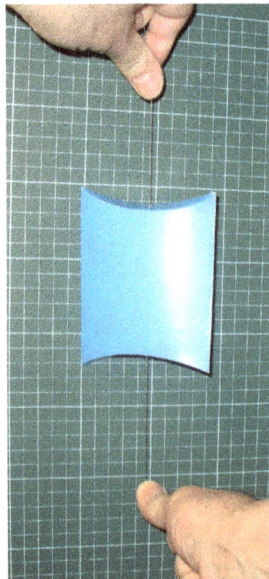

**Figure 5.13**

- Transparent adhesive tape
- A piece of any light straw with a diameter of 3–4 mm

Method:

1. On the reverse side of half a sheet of coloured craft paper, draw an outline for the future toy, as shown in Figure 5.14.
2. To facilitate future bends of the craft paper, push with a ballpoint pen with sufficient force along the dotted lines shown in Figure 5.15, pierce four holes with a diameter of 4–5 mm, and adhere a piece of straw with tape, as shown in Figure 5.15.
3. Bend the card piece along the dotted lines, as shown in Figure 5.16.
4. Thread the thread through the punctured holes and the piece of straw, as shown in Figure 5.17. To prevent the thread from subsequently jumping out of the toy, it is worth tying small pieces of craft paper at its ends.
5. Bend the craft paper piece into a box and adhere it with tape along the side seam, as shown in Figure 5.18.

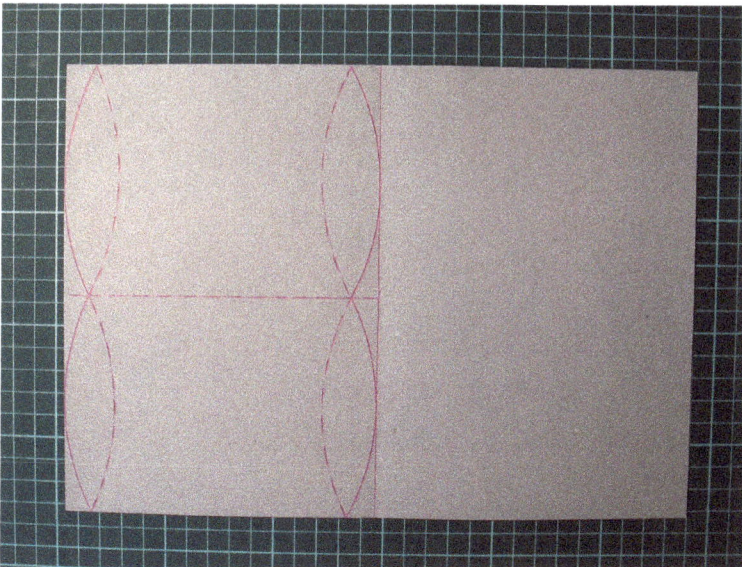

**Figure 5.14**

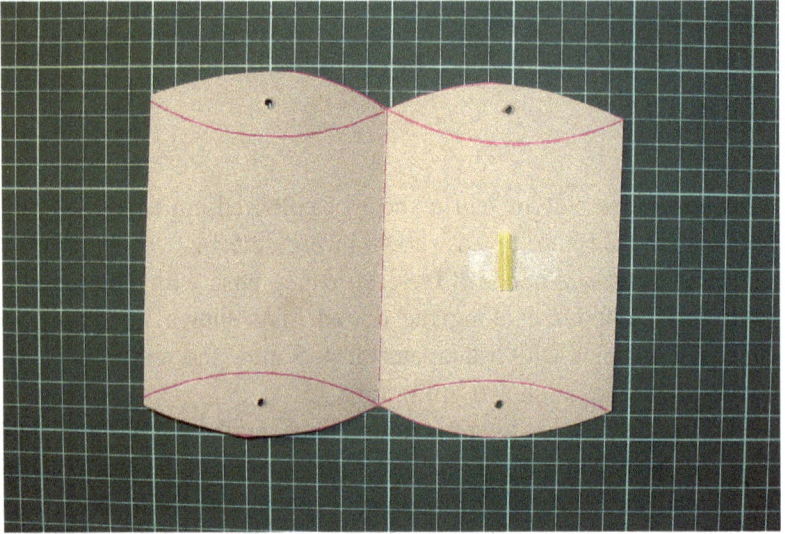

**Figure 5.15**

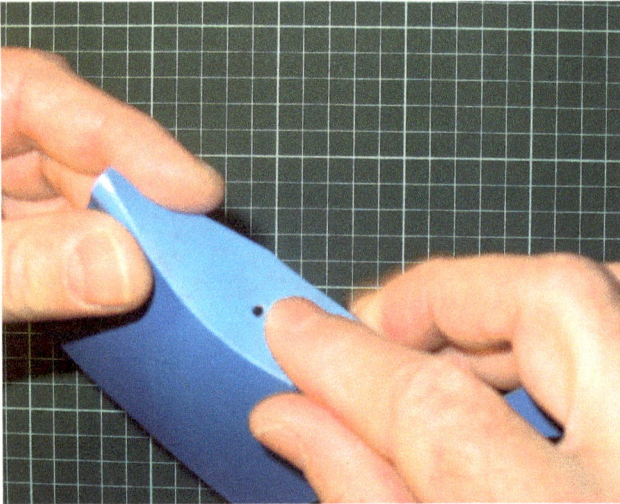

**Figure 5.16**

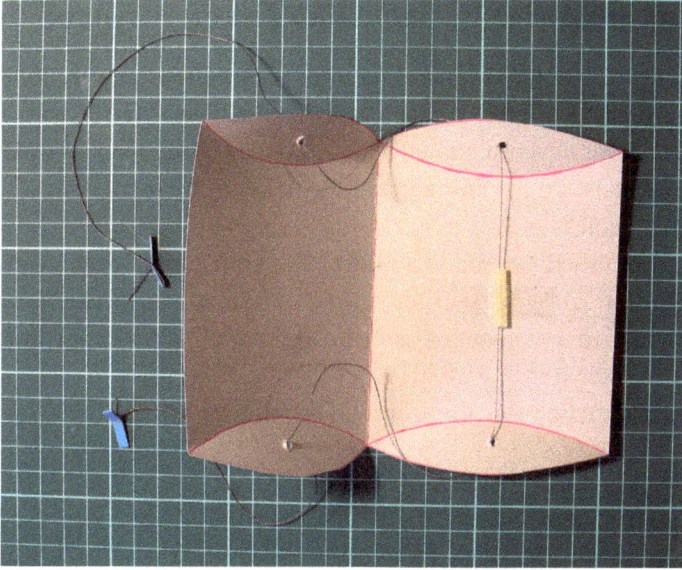

**Figure 5.17**

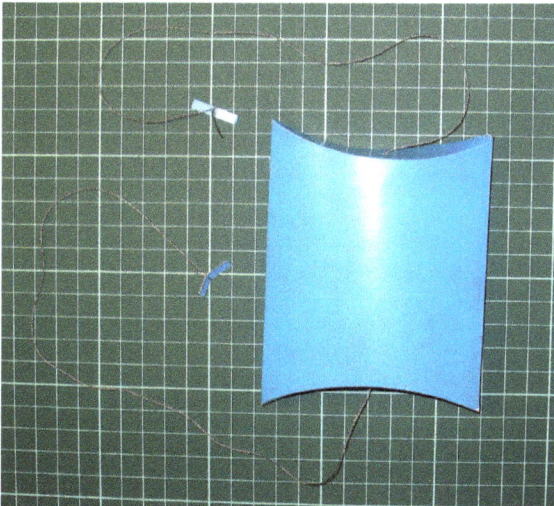

**Figure 5.18**

## 5.4 Non-motorised Cord Model of an Airplane

Despite the rapid development of radio-controlled aircraft modelling, cordless manned aircraft models still have enthusiasts for this type of modelling. Classic cord models are equipped with special small-sized internal combustion engines and can be made to fly in a circle when tied to two thin cables.

This simplified version is a glider with craft paper wings with a density of at least 320 g/m$^2$, with its fuselage made of straw. The model flies in a circle due to a moment of centrifugal force of inertia being created by a thread, as shown in Figure 5.19.

For constructing this toy, you will require:

- A sheet of A4 coloured craft paper
- Thread
- Transparent adhesive tape
- A straw with a diameter of 8 mm
- Three or four paper clips

Method:

1. Wing pieces are cut out of the craft paper cover, as shown in Figure 5.20.

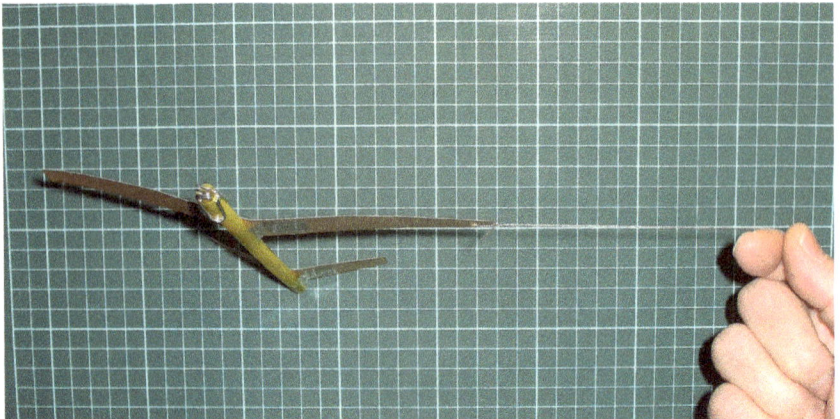

**Figure 5.19**

2.   The pieces of the wings are adhered with strips of tape to the straw, as shown in Figure 5.21.
3.   At the end of one of the wings, a hole is pierced for a thread, as shown in Figure 5.22.

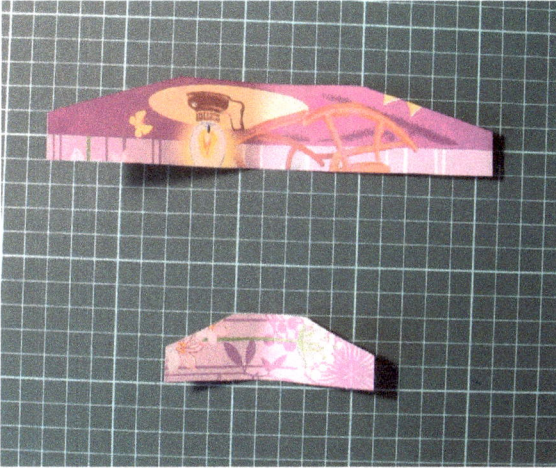

**Figure 5.20**

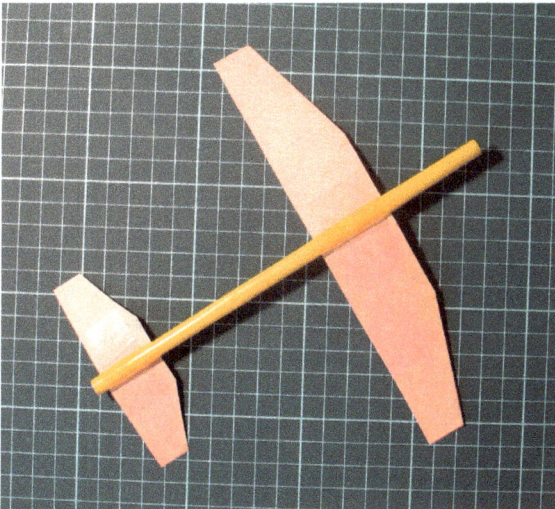

**Figure 5.21**

4.   A thread is threaded through the hole and tied to the wing, as shown
     in Figure 5.23.

**Figure 5.22**

**Figure 5.23**

**Figure 5.24**

5.  For balancing, paper clips may be inserted into the straw, as shown in Figure 5.24.

In flight, the toy is powered by the rotation of the hand holding a metre or so of thread. Flight is achieved through proper balancing by an appropriate number of paper clips, one can also bend the trailing edge of the tail to alter direction.

## 5.5  Sailboat (Made from foam)

The body of the model of a three-masted sailboat is made of three fragments of ceiling foam tiles with a thickness of 5 mm, connected by means of tape. When launched on an open pond, the sailboat model holds steady on small waves and catches the wind well. The general view of the sailboat model is shown in Figure 5.25.

For constructing this toy, you will require:

- One foam ceiling tile
- A sheet of A4 coloured craft paper
- Electrical insulation tape
- Three straws with a diameter of 5 mm
- Scissors

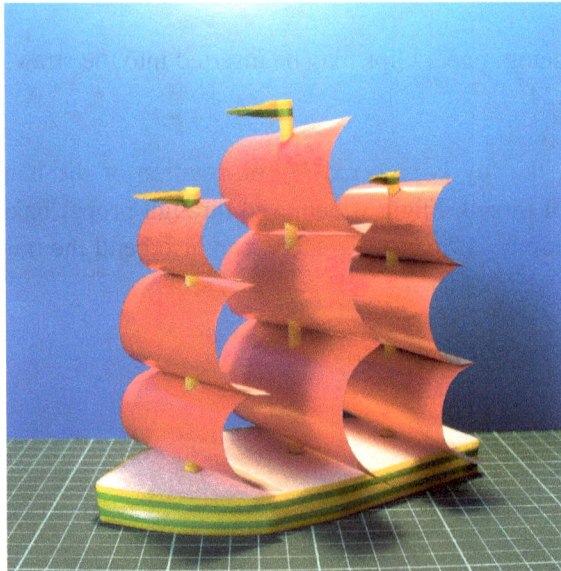

**Figure 5.25**

Method:

1. On three fragments of foam tiles, the contours of the future hull of a sailboat with a length of about 10–15 cm are marked, as shown in Figure 5.26.
2. Cut the three pieces of the model body out using scissors, as shown in Figure 5.27.
3. All three pieces are connected together using tape, as shown in Figure 5.28.
4. The three straws are shortened to approximately the size shown in Figure 5.29. At one end of each of them, four to five layers of tape are wound, as shown in Figure 5.29.
5. With a sharpened pencil, the models are pierced and expanded to a diameter of approximately 7–8 mm. As a basis for the holes for three masts, refer to Figure 5.30.
6. The three masts should be tightly inserted into the model body, as shown in Figure 5.31.

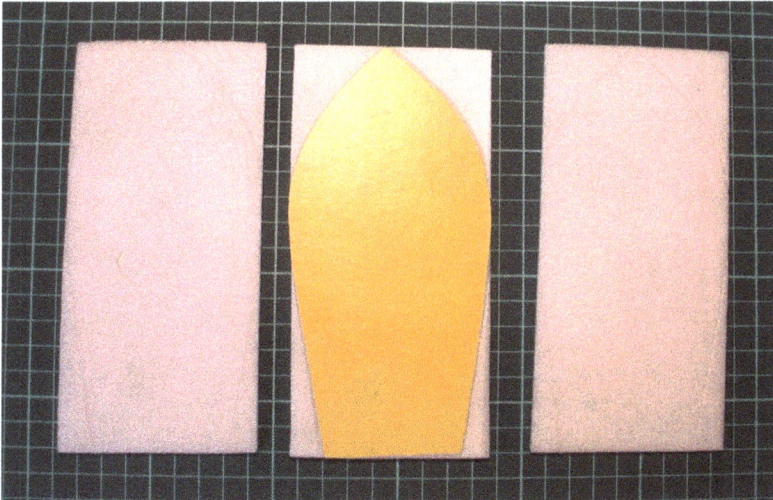

**Figure 5.26**

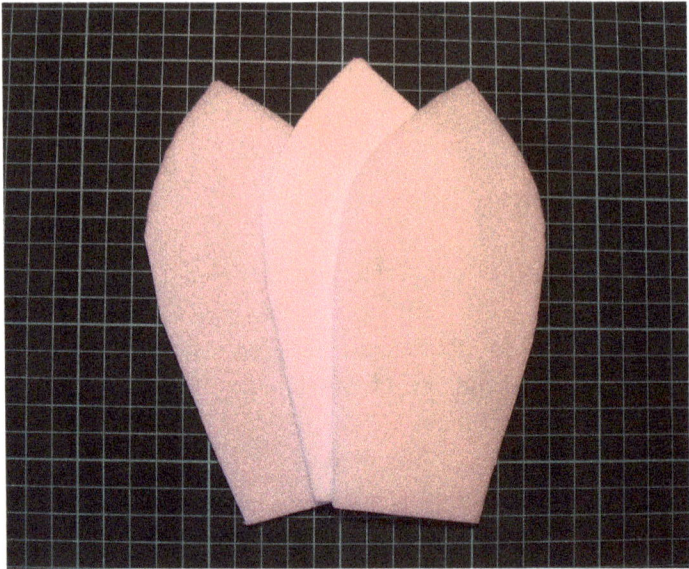

**Figure 5.27**

**Figure 5.28**

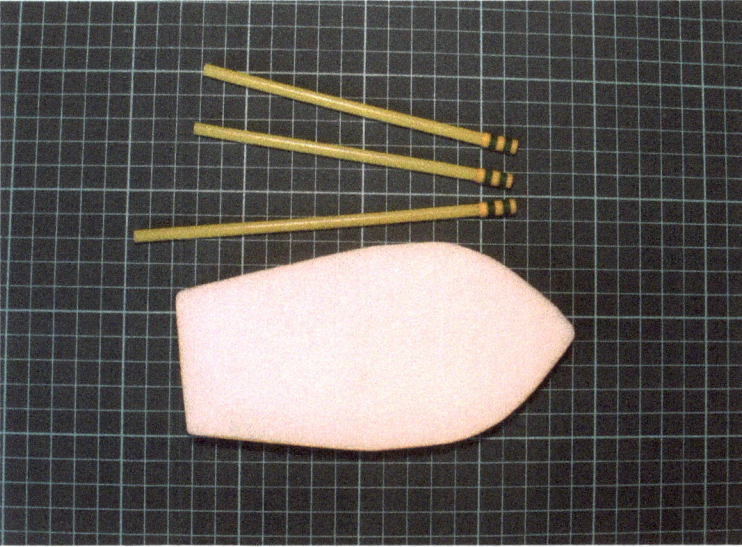

**Figure 5.29**

**Figure 5.30**

**Figure 5.31**

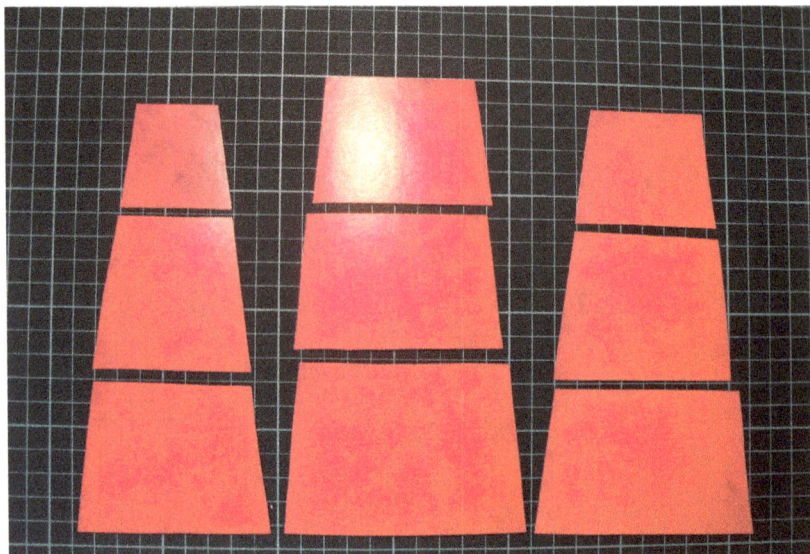

**Figure 5.32**

**Figure 5.33**

7. From a sheet of A4 coloured craft paper, sail pieces are cut out, as shown in Figure 5.32.
8. By curling the cut craft paper over the edge of the table, the sail pieces are given some curvature, as shown in Figure 5.33. After that, holes for the masts are made with a clerical hole punch, also as can be seen in Figure 5.33
9. **At the final stage, the sails are put on the masts and fixed there with the help of stylised flags made of tape, as shown in Figure 5.25.**

## 5.6  Ballista (Made from the bottom of a plastic bottle)

The principle of operation of the oldest torsion throwing machines is demonstrated by a homemade ballista made from a plastic bottle, as shown in Figure 5.34.

For constructing this toy, you will require:

- Plastic bottle
- Nylon cord with a diameter of 2–4 mm
- Bamboo skewer

Method:

1. Cut off the bottom of the bottle with a height of about 7–8 cm, as shown in Figure 5.35.
2. Pierce two holes in the bottom, expanding them to a diameter of 4–5 mm, as shown in Figure 5.36.
3. Thread a piece of cord through two holes and fix it with two pieces of a skewer and a strong knot, as shown in Figure 5.37.

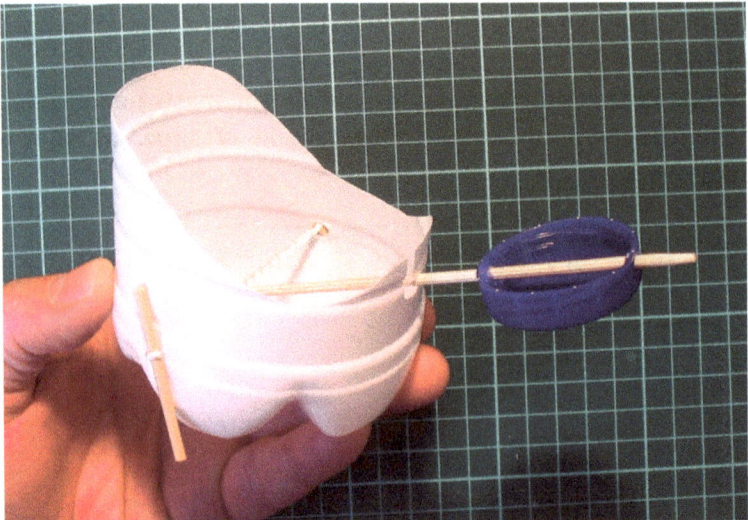

**Figure 5.34**

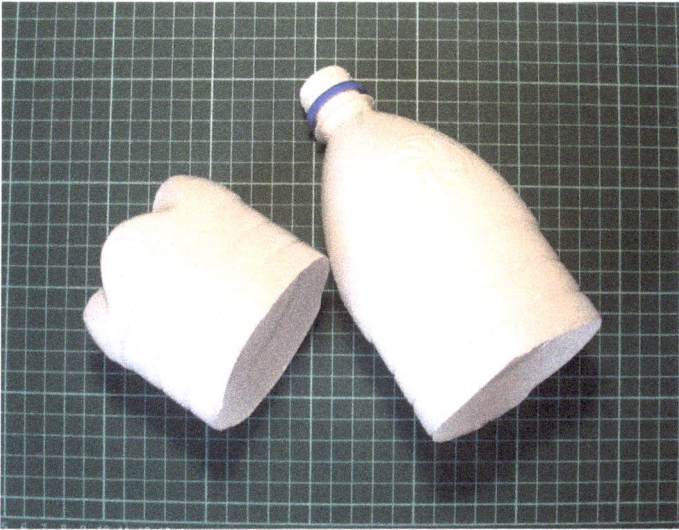

**Figure 5.35**

**Figure 5.36**

4. To be able to fix the cocked ballista before release, cut off the bottom of the bottle, as shown in Figure 5.38.
5. Make a throwing lever by piercing two holes in the bottle top and inserting a skewer, shortened to 10–12 cm, into those holes, see Figure 5.39.

**Figure 5.37**

**Figure 5.38**

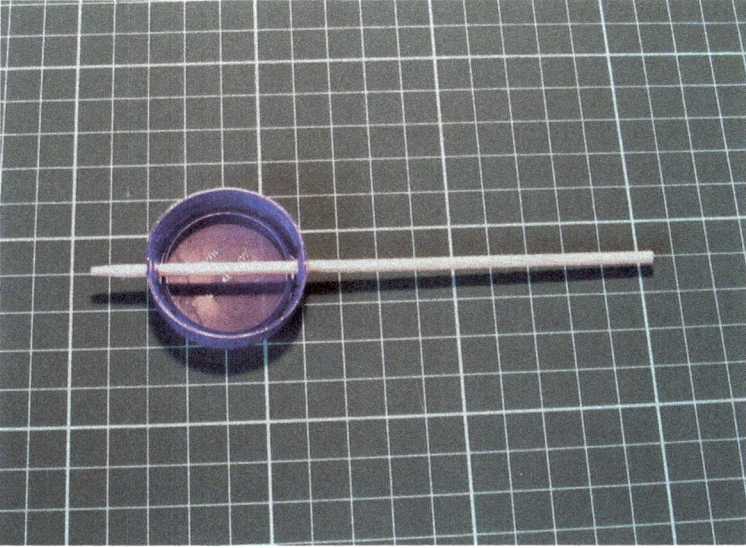

**Figure 5.39**

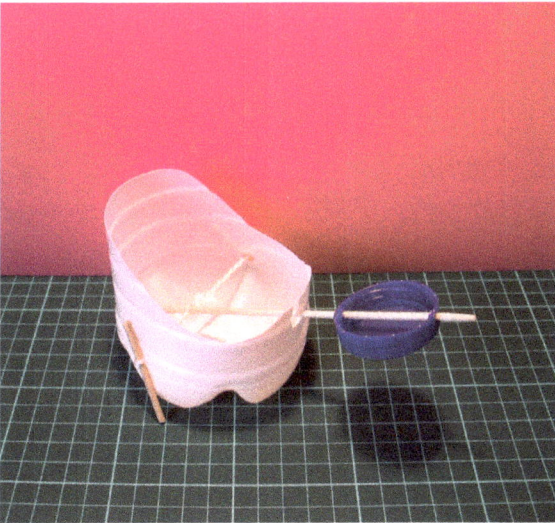

**Figure 5.40**

6.  Insert the bottle top on the skewer into the bottom of the bottle with the threaded cord, as shown in Figure 5.40.

To make a shot from a ballista, you need to put a small light weight in the bottle top on the lever, pull back and fix the skewer in the rut which was cut from out of the bottom of the bottle, and then, at the right moment, push the skewer out of the rut. The power of the shot can be adjusted by turning the pieces of skewer that hold the nylon cord. Do not twist them too much, as you could deform the bottom of the bottle.

## 5.7 Wound Rubber Powered Submarine (Made from a plastic bottle with a sugar cube surfacing machine)

This homemade toy (as seen in Figure 5.41) has multiple interesting functions. In addition to the wound rubber motor, it can float independently about a minute after diving. The toy works as follows:

- Fill a plastic bottle with water so that it barely keeps afloat and seal its top hermetically;
- Tie a thread with two free ends about 15 cm long to a small weight that can drag the bottle to the bottom of the thread;
- Tie each of the free ends of the thread tightly around the bottle to a piece of sugar so that the weight is at the bottom of the bottle;
- Let the toy float freely then watch it sink under the water;
- Wait a little while until the sugar dissolves, the thread is released, the cargo falls to the bottom, and the bottle floats to the surface.

For constructing this toy, you will require:

- Plastic bottle
- Thread

**Figure 5.41**

- A small piece of plastic from the bottle
- Two toothpicks
- Electrical insulation tape
- Stationery elastic band
- Lump of sugar

Method:

1. Cut off a 10 cm piece of straw with a diameter of 8 mm, so that the stationery elastic band is put on it with a slight tension, as shown in Figure 5.42.
2. For rigidity, wind the electrical insulation tape around the edges of the cut straw seven to eight times, and make two incisions on one end, as shown in Figure 5.43.
3. Mark a fragment of a plastic bottle for the future water screw and two elements of the sliding bearing, as shown in Figure 5.44.
4. Cut out the pieces of the water screw and the two elements of the sliding bearing and pierce them, expanding the holes to about 3 mm in diameter. For bending the water screw, refer to Figure 5.45.

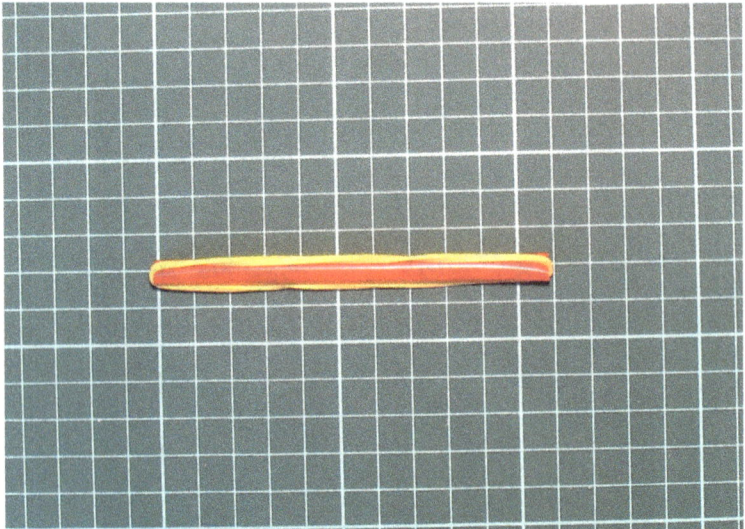

**Figure 5.42**

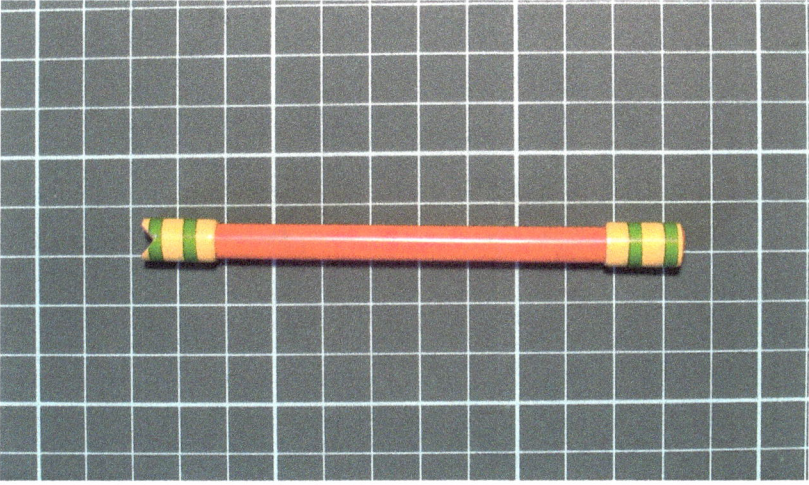

**Figure 5.43**

**Figure 5.44**

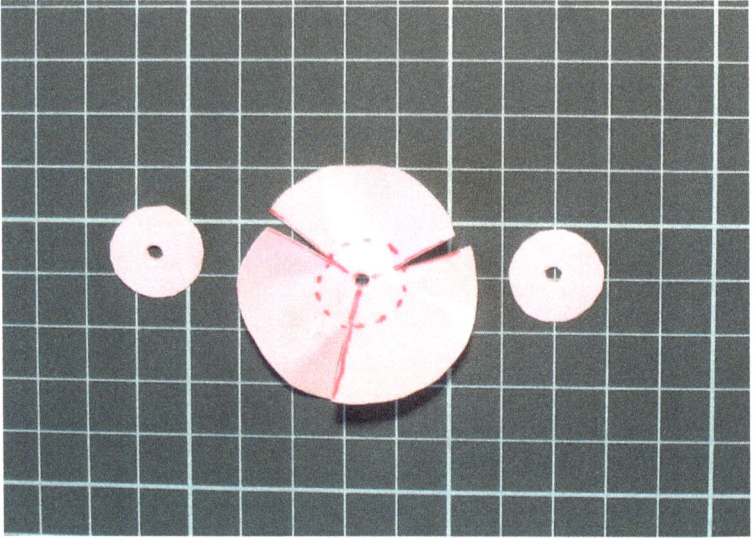

**Figure 5.45**

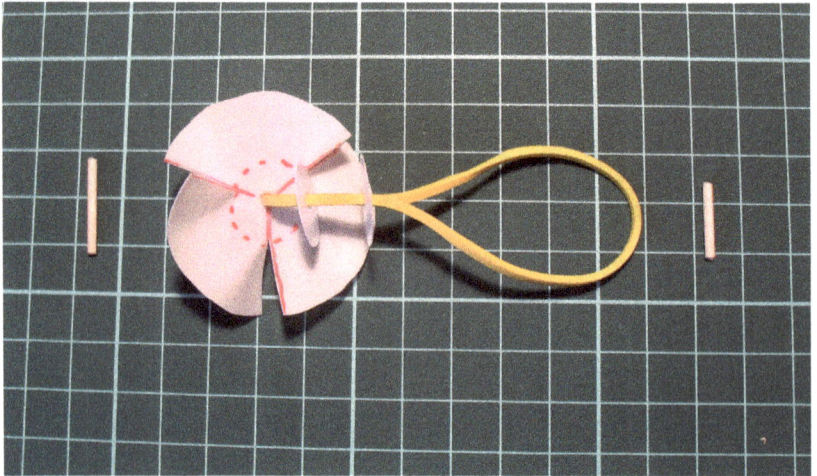

**Figure 5.46**

5.  Cut two 15 mm pieces from a toothpick, threading a stationery elastic
    band through the water screw and two elements of the sliding bearing,
    as shown in Figure 5.46.

6. Bend the hook from the paper clip to tighten the rubber motor and prepare the elastic band before installing it into the tube by pulling the water screw, the sliding bearing elements, and a piece of toothpick, as shown in Figure 5.47.

7. Insert an elastic band with a screw and a sliding bearing into the tube, pull it with a hook from a paper clip and fix it with a piece of a toothpick, as shown in Figure 5.48.

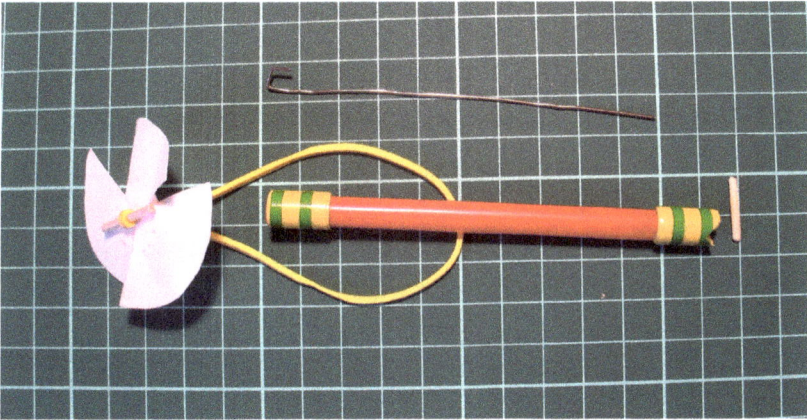

**Figure 5.47**

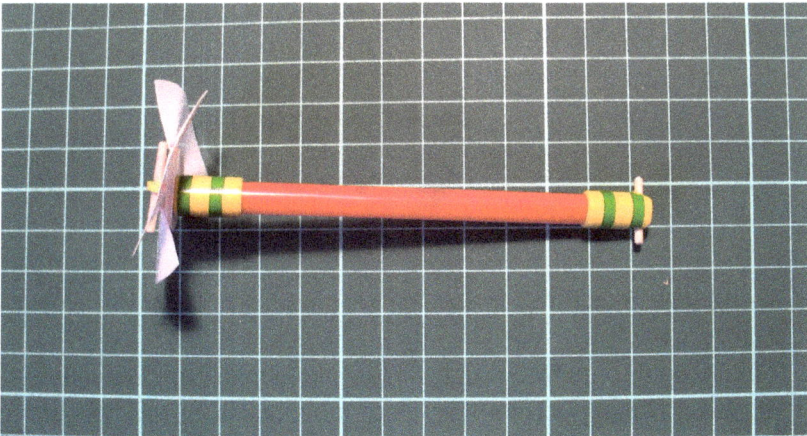

**Figure 5.48**

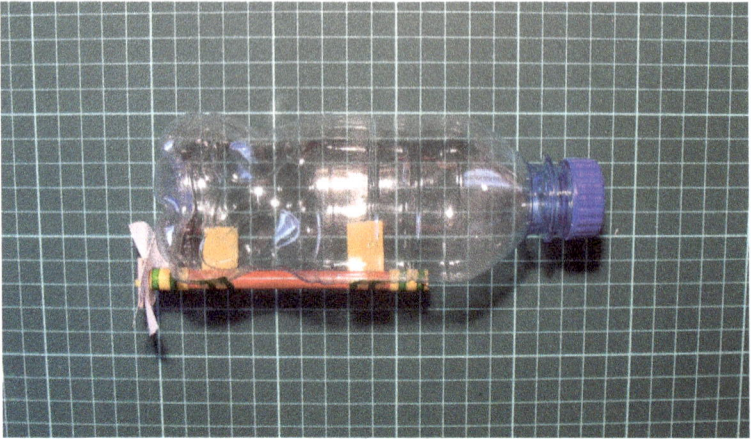

**Figure 5.49**

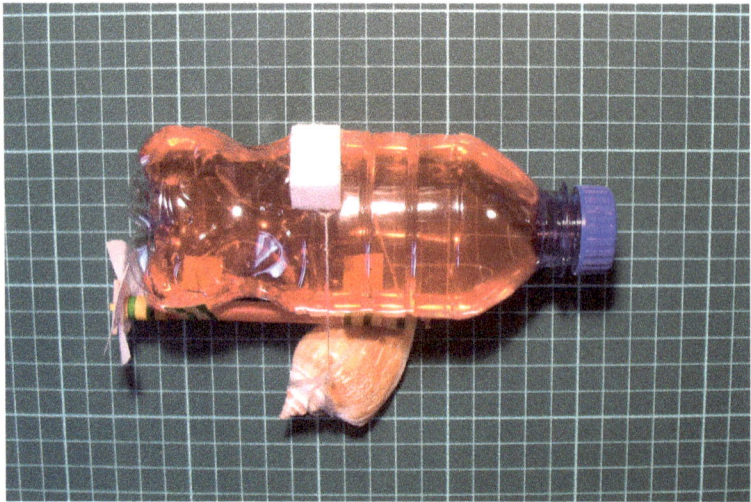

**Figure 5.50**

8.  Using tape, attach the rubber motor to a plastic bottle, as shown in Figure 5.49.
9.  Fill the bottle with water (tinted in the photo to improve visualisation) so that it is barely kept afloat, close it hermetically with a bottle top and attach an additional load by tying the two ends of the thread about a piece of sugar, as shown in Figure 5.50.

## 5.8  Pneumatic Arrow (Made from a plastic bottle)

With sudden pressure applied to the bottle, one can quite accurately shoot a straw to a distance of 2–3 m.

For constructing this toy, you will require:

- Plastic bottle
- Straws with diameters of 8 mm and 4 mm
- Copper wire with a diameter of 1.5–2 mm, of approximately 20 cm length
- Hot melt glue or similar adhesive

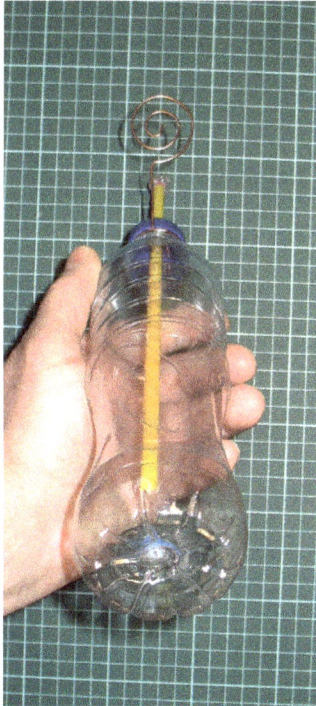

**Figure 5.51**

Method:

1.  In the plastic bottle top, make a hole with a diameter of slightly more than 8 mm, as shown in Figure 5.52.
2.  Insert a straw with a diameter of 8 mm into the bottle top, sealing it tightly with hot glue. The smaller straw will be hermetically sealed at one end with hot glue as shown in Figure 5.53.
3.  Tightly tighten the bottle top, inserting a thin straw into a thick one. The arrow is ready, as shown in Figure 5.54.

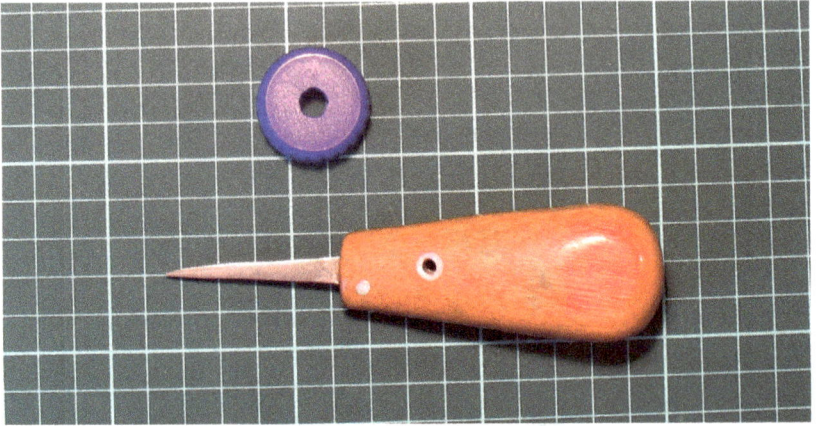

**Figure 5.52**

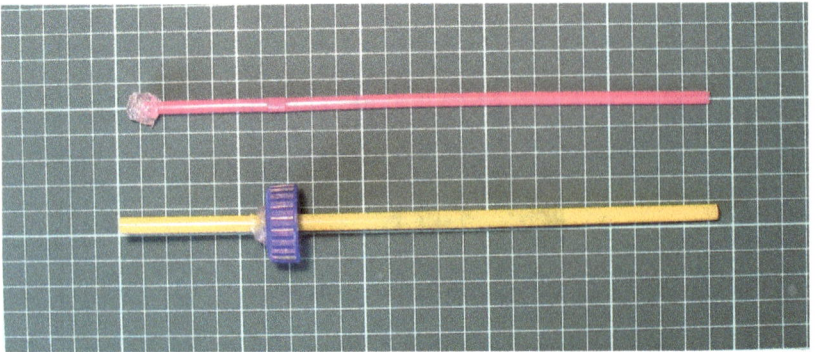

**Figure 5.53**

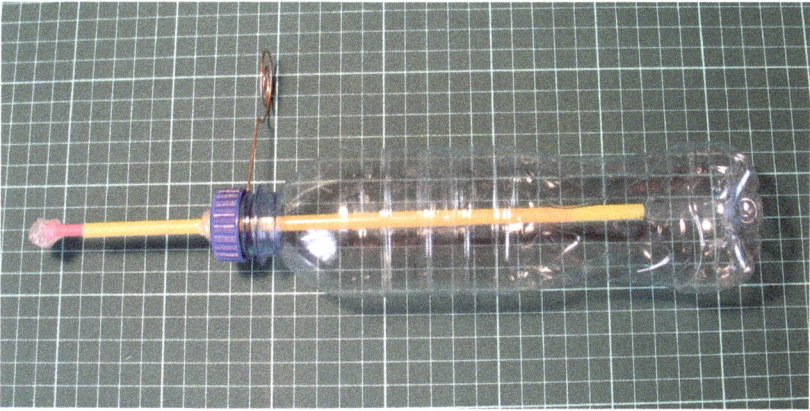

**Figure 5.54**

## 5.9  Self-made Cartesian Diver

There are a lot of design options for a homemade Cartesian diver, but perhaps the simplest of them is shown in Figure 5.55.

For constructing this toy, you will require:

- Plastic bottle
- A straw with a diameter of 8 mm
- Three to five paper clips
- Electrical insulation tape

Method:

1. A piece of about 8 cm is cut off from the straw. From one of the ends, 1 cm of the straw is bent back on itself and, when pressed to the long side, it is fixed with tape, as shown in Figure 5.56.
2. A hole is pierced from the other edge of the straw and three to five paper clips are inserted into it, as shown in Figure 5.57.

**Figure 5.55**

3. The diver descends into a bottle filled to the top with water and is hermetically sealed within by the bottle top. When pressed by hand, the diver should sink, and after the pressure is released, it should float. If necessary, the weight of the diver can be adjusted by adding or subtracting paper clips.

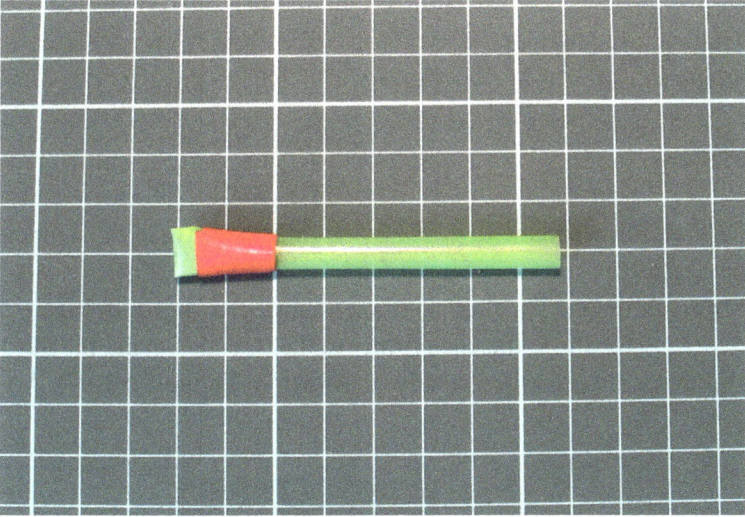

**Figure 5.56**

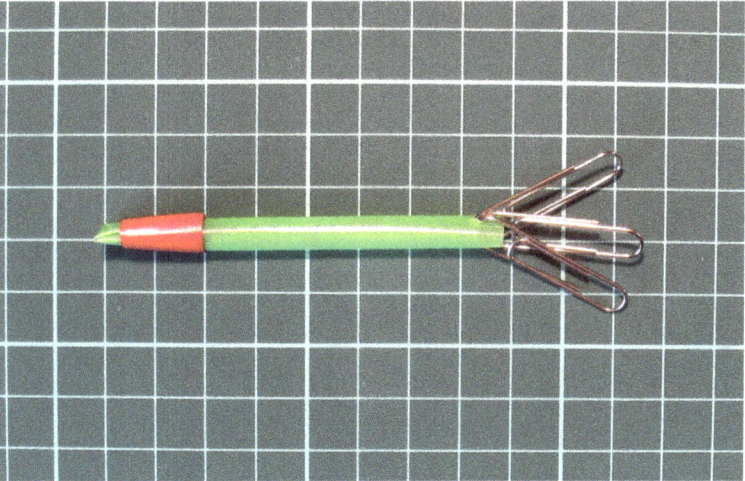

**Figure 5.57**

## 5.10 Astrojax (Made from ping-pong balls)

Astrojax are not always easy to find for sale — their popularity varies. However, it is very easy to make it yourself from three weights and a thread, for example, from three ping-pong balls, as shown in Figure 5.58

For constructing this toy, you will require:

- Three ping-pong balls
- Nylon thread

Method:

1. Drill all three balls with a drill bit slightly larger than the diameter of the thread, as shown in Figure 5.59.
2. String all three balls alternately on a thread, as shown in Figure 5.60.
3. Tie the two balls at the ends firmly to the thread, leaving the third one sliding freely. The toy is ready, as shown in Figure 5.58.

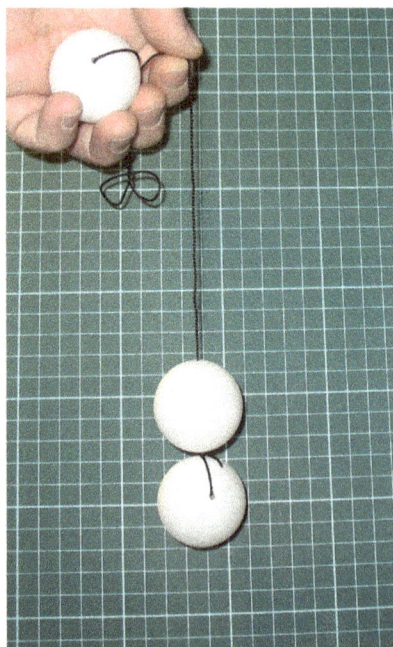

**Figure 5.58**

**Figure 5.59**

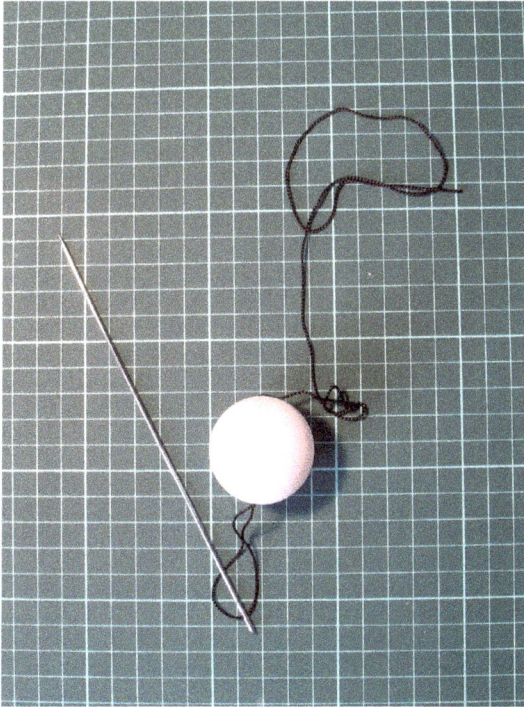

**Figure 5.60**

## 5.11  Hydraulic Press (Made from medical syringes)

The principle of operation of hydraulic machines is based on the directly proportional ratio of the force on the piston and its area. This is easy to understand when playing with the simplest device of two medical syringes connected by a small piece of plastic tube from a medical dropper, as shown in Figure 5.61.

For constructing this toy, you will require:

- A couple of medical syringes of various volumes
- A piece of plastic tube from a disposable medical dropper about 15 cm in length

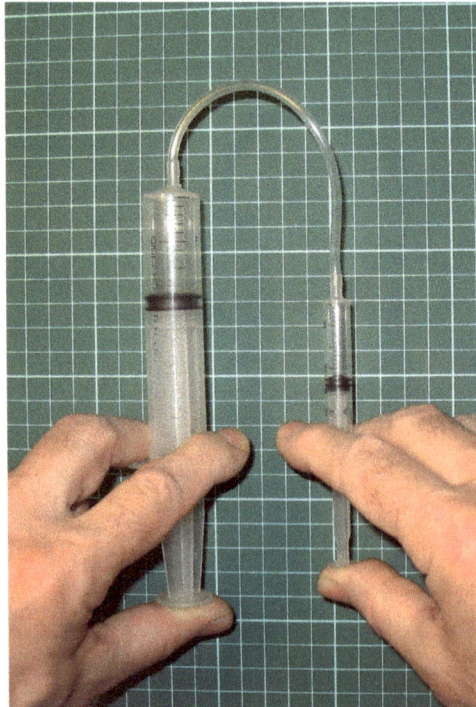

**Figure 5.61**

It is very simple to make the device. One must connect the smaller of the two syringes filled with water with a plastic tube to the large syringe, as show in Figure 5.62.

Replacing the smaller of the syringes with an intermediary size of syringe, it is easy to feel the difference in the forces applied to the pistons of the syringes compared to the degree of their movement.

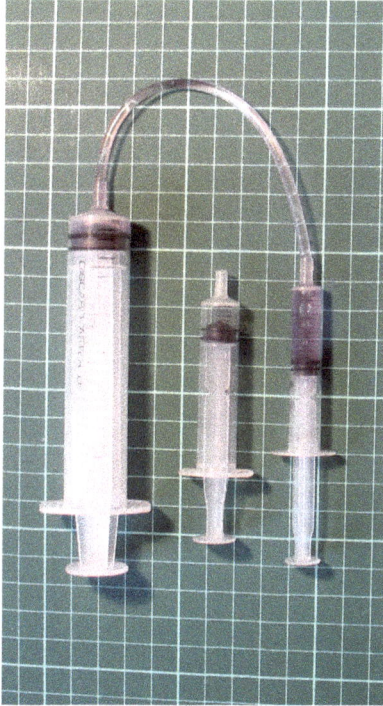

**Figure 5.62**

## 5.12  Demonstration of Pascal's Law (Displayed with the help of a pipette and a syringe)

Pascal's law, which states that the pressure in liquids and gases is distributed equally in all directions with the help of a syringe and a pipette, is clearly demonstrated by a simple homemade toy shown in Figure 5.63.

For constructing this toy, you will require:

- Medical syringe with a volume of 20 Ml
- Plastic pipette
- Plastic tube adapter from a disposable medical dropper

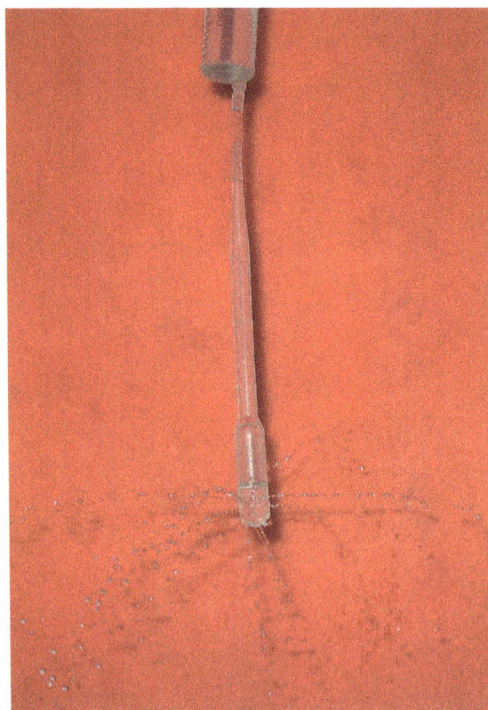

**Figure 5.63**

Method:

1. Repeatedly pierce the pipette with an awl in different directions, as shown in Figure 5.64.
2. Using a piece of plastic tube from the dropper, connect the syringe filled with water and the punctured pipette, as shown in Figure 5.65.

If you press on the plunger of the syringe, water will pour out in jets directed according to Pascal's law in all directions, as shown in Figure 5.63.

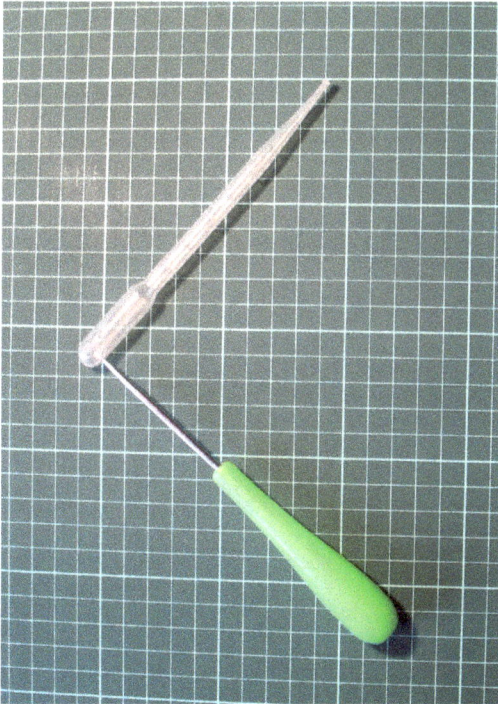

**Figure 5.64**

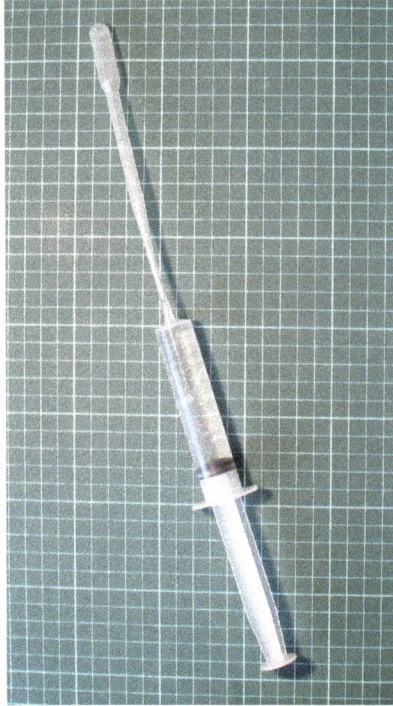

**Figure 5.65**

# 5.13 Air Gun (Made from a plastic jar)

For this pneumatic toy, any plastic jar with a volume of about 500 mL with a removable lid that closes tightly enough is suitable. If you cut a small round hole in the bottom of the jar, then by hitting the lid with a click of your finger, you can create a pressure pulse in such a way that the air escaping through the hole in the bottom forms a toroidal seal that can extinguish the candle from roughly 15 cm away. Such a toy is shown in Figure 5.66.

For constructing this toy, you will require only one plastic jar with a lid with a volume of half a litre.

It is very simple to make a toy, you just need to cut a round hole in the bottom of the jar with a diameter of about 2 cm, as shown in Figure 5.67.

**Figure 5.66**

**Figure 5.67**

## 5.14  Sounding Turbulence

To see how the air flow moves in a disorderly vortex while making noise, you can use a simple homemade whistle from a straw, as shown in Figure 5.68.

Method:

1.  Cut off a 7 cm piece from a straw, as shown in Figure 5.69

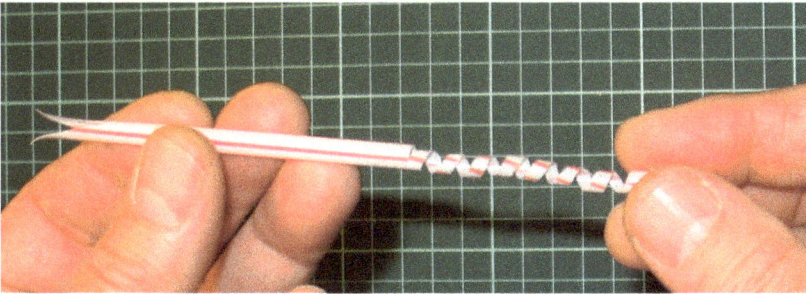

**Figure 5.68**

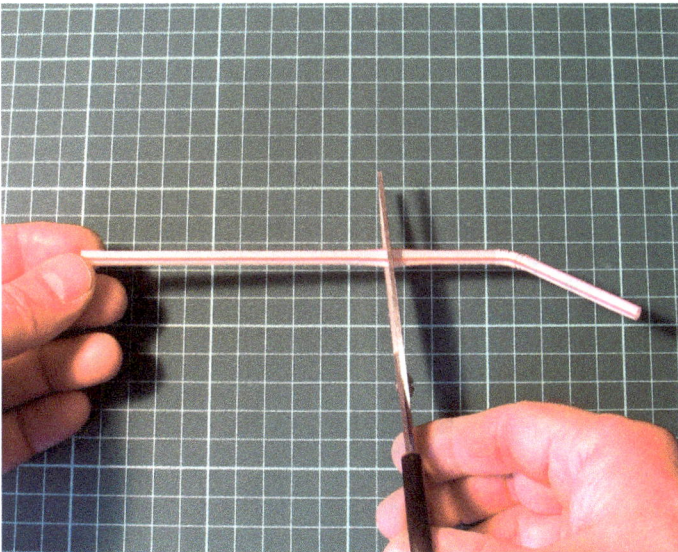

**Figure 5.69**

2. Cut one of the ends of the straw from both sides at an angle of about 15°, as shown in Figure 5.70.

3. After holding and pressing the scissors to cut the pointed edges of the straw, deform them slightly, as shown in Figure 5.71. The result will be a squeaking sound on use, coming from two vibrating elastic plates converging and diverging with the sound vibrations of the air, which are present in any musical wind instrument (whether it is the double reed of the bassoon or the lips of the trombonist).

4. From the other edge of the straw, gradually cut a couple of times, as shown in Figure 5.72, to form a spiral 2–3 cm in length, also shown in Figure 5.72.

To make the toy whistle, you should take the straw between your lips so that the edges which will whistle are 3 cm from the front edges of your

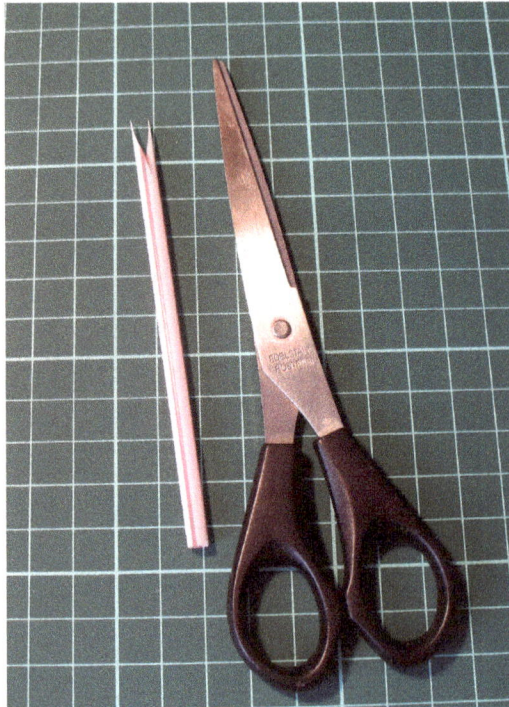

**Figure 5.70**

teeth. It is necessary to blow into the straw with sufficient pressure, but if one does then the spiral on the opposite edge of the straw will clearly demonstrate how the air moves in a turbulent flow.

**Figure 5.71**

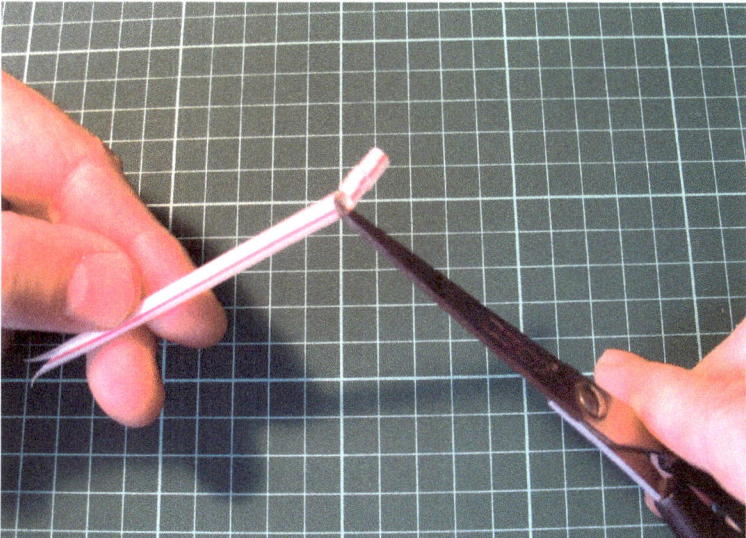

**Figure 5.72**

# Index